THE WORLD IN A BOX

THE WORLD IN A BOX

The Story of an Eighteenth-Century Picture Encyclopedia

Anke te Heesen

Translated by Ann M. Hentschel

The University of Chicago Press
Chicago and London

Anke te Heesen is a research associate at the
Max Planck Institute for the History of Science in Berlin.

Typeset by Ann M. Hentschel

This work was originally published as
Der Weltkasten: die Geschichte einer Bildenzyklopädie aus dem 18. Jahrhundert, © Wallstein Verlag 1997.

The University of Chicago Press, Chicago 60637
The University of Chicago Press, Ltd., London

Printed in the United States of America

11 10 09 08 07 06 05 04 03 02   1 2 3 4 5

ISBN: 0-226-32286-6 (cloth)
ISBN: 0-226-32287-4 (paper)

Library of Congress Cataloging-in-Publication Data

Heesen, Anke te
[Weltkasten. English]
The world in a box : the story of an eighteenth-century picture encyclopedia / Anke te Heesen ; translated by Ann M. Hentschel.
p. cm.
Includes bibliographical references and index.
ISBN 0-226-32286-6 (alk. paper) — ISBN 0-226-32287-4 (pbk. : alk. paper)
1. Weltkasten—History. 2. Children's encyclopedias and dictionaries, German—History and criticism. 3. Encyclopedias and dictionaries, German—History and criticism. 4. Picture dictionaries, German—History. 5. Enlightenment—Germany—History—18th century. 6. Material culture—Germany—History—18th century. 7. Visual communication—Germany—History—18th century. 8. Germany—Intellectual life—18th century. I. Title.

AG27.W433 H44 2002
033′.1—dc21

2001052763

♾ The paper used in this publication meets the minimum requirements of the American National Standard for Information Sciences—Permanence of Paper for Printed Library Materials, ANSI Z39.48-1992.

CONTENTS

ILLUSTRATIONS

Copperplate illustrations were frequently bound at the back of eighteenth-century works and, as with Stoy's Picture Academy, many were sold separately as well in complete sets. In an adaptation of the historical formal element of the picture encyclopedia, this monograph reproduces individual engravings (labeled as figures) where they are discussed in the text and assembles thematic picture series or tableaux (plates) after the chapters where they are mentioned. The topics of these illustrations have been incorporated in the index to facilitate reference between image and text.

Figures

Tableau A *(follows page 98)*

Tableau B *(follows page 126)*

Tableau C *(follows page 154)*

Tableau D *(follows page 186)*

ACKNOWLEDGMENTS

Writing is a solitary business. It is, however, part of a collective activity, where computer files are assembled and index cards collected in a box, where people gather ideas, moral support, enthusiasm, and chocolate. So I owe many thanks to many people.

First and foremost I would like to thank Rudolf Prinz zur Lippe and Konrad Wünsche for their critical discussions and generous support. Theodor Brüggemann and Gertrud Strobach kindly made their children's book collections available to me. I am indebted to the librarians and archivists at numerous institutions for their indispensable assistance, particularly at the Stadtbibliothek Nürnberg and the Niedersächsische Staats- und Universitätsbibliothek Göttingen. I would like to thank both these institutions and the Staatsbibliothek zu Berlin Preußischer Kulturbesitz for generously providing the illustrations. While working on this book, I was supported by a dissertation grant from the Konrad-Adenauer-Stiftung.

Various friends and colleagues gave me both moral and intellectual support, offering me useful pointers along the way. I would like to thank Michael Becker, William Clark, Alix Cooper, Elisabeth Fisch, Martin Gierl, Eva Glaum, Andreas Hartmann, Doris Kaufmann, Heidrun Ludwig, Christine Reuter, Hans-Jörg Rheinberger, Wiebke Richert, Skuli Sigurdsson, Bettina Wahrig-Schmidt, and Silke Wenk. Special thanks are due to Michael Hagner for his inexhaustible solidarity throughout different stages of completeness of manuscript and author.

For the English version of this book I would like to thank Ann Hentschel, not only for the careful translation, but also for the checking of bibliographic references, which was of benefit to the book as a whole. Without Emma Spary I could not have kept the spirit of the *Weltkasten* alive among the vagaries of language and argumentation. From the beginning, she gave support, encouragement, and wonderful intellectual discussions. Abigail Lustig and Simon Werrett helped me as friends and colleagues. They kindly read through an earlier version of this English translation and made invaluable suggestions.

Thanks to the accuracy of the copyeditor, Joann Hoy, errors were corrected and oversights caught. Christine Schwab was the production editor, Dennis Anderson was responsible for the design, and Siobhan Drummond for production control. Finally, I have to thank Susan Abrams for her insistence at the right moments, encouragement in numerous situations, and above all, patience.

A Note about This Translation

Since completion of the German version of this book in 1996, a whole range of new works on the history of collecting and the history of images has appeared. Where the recent literature addressed specific issues raised in this book, I made the necessary alterations in the footnotes or references. Nevertheless, with the exception of the introduction, the English edition remains essentially the same as the German original.

THE WORLD IN A BOX

INTRODUCTION

"A Storehouse of the Finest Materials"

Figure 1. Storage box of Stoy's Picture Academy
(photograph by the author)

As the cardboard-backed illustrations are pulled out, their contents come into view: Romulus appears beside a mason, herrings follow a wolf, and the Greek Muses meet Abraham and his son Isaac—a bewildering array of images and themes. They are stored inside a box measuring seventeen by twelve inches. It is divided into compartments of different sizes in which 468 copperplate engravings are filed. A whole world in a box.

The creator of this Picture Academy for the Young *(Bilder-Akademie für die Jugend),* the German theologian Johann Siegmund Stoy, sought to design an encyclopedic propaedeutic similar to that described by his predecessor Johann Amos Comenius toward the close of the seventeenth century. "Let me guide you through all things," he invites his readers in the introduction to his famous and groundbreaking *Orbis sensualium pictus quadrilingus,* "I will show you everything; I will name you everything" (Comenius 1685). Comenius was one of the first to seek to supply *all* necessary and fundamental knowledge for young people by means of a comprehensive book and its pictures. In designing his Picture Academy for the Young, Stoy offered something along the same lines. In the works of Comenius and Stoy, "everything" was systematized, organized, converted into word and image, compartmentalized. "Everything" meant the whole world. For, as Stoy said, the world is "the earth on which mankind lives, and the heavens which we see above us" (1780–84, 1). To order the plenitude and diversity of the objects of this world, to comprehend their totality in a glance—these are the purposes of Stoy's box with its picture cards.

The Grimms's dictionary defines the German word for box, *Kasten,* as an "*arca, cista,* usually made from wood, but also as carton, iron strongbox for money . . . ; of related meaning to box are crate, cupboard, coffer, chest, drawer" (1984, 11:264).[1] The word *Kasten* was used in household affairs and the transport of goods; the book trade had the term *Zettelkasten,* or card-index tray, and the "typesetter's frame at a printers" (p. 265). World and box, content and ordering medium are thus bound together. If neither Stoy nor Grimm ever used the expression *Weltkasten,* literally "world box," it is nevertheless valuable for describing the central, material means of ordering as well as the encyclopedic arrangement of the contents: the "world in a box" should thus be understood in a literal sense.

The box appeared, in the first instance, in the shape of a book. This work, entitled Picture Academy for the Young, consists of one folio volume of illustrations, with fifty-two copperplate engravings, and two octavo volumes of text. The author recommended that some sets of these engravings be "cut up according to subject" (*Allgemeine Bibliothek* 1782, 464), with the resulting individual prints to be glued onto cardboard and then sorted in a box specially made for the purpose.

The author, Johann Siegmund Stoy, was a priest in a little Protestant parish near Nuremberg until he settled on the publication of the Picture Academy as his principal occupation. Having resigned his ministry, he moved to the city and became a "professor of pedagogy," planning an educational institute that never materialized. Instead, Stoy set up an "education business" *(Edukationshandel)* or "pedagogical cabinet" that offered visual teaching material for children.

There was a wide market for such teaching materials at the end of the eighteenth century. In accordance with the ideals of the Enlightenment, a child was considered capable of development and gradual perfection. Above all, his or

1. The principal definition for *box* in *The Oxford English Dictionary* (2d ed., 2:461) reads, "A case or receptacle usually having a lid; *(a)* orig. applied to a small receptacle of any material for drugs, ointments, or valuables; *(b)* gradually extended (since 1700) to include cases of larger size, made to hold merchandise and personal property; but (unless otherwise specified) understood to be four-sided and of wood."

her capacity for sensation—the foundation of all further knowledge—was to be shaped by means of appropriate resources. It was for this purpose that "sensational aids" *(Versinnlichungsmittel)* were devised, in other words, objects such as pictures, models, and books, which developed the child's sensibility. Such appropriated forms of knowledge were supposed to enable children to acquire abilities that would encourage them to independent thought and judgment, and enable them to become useful members of society. Within the context of the educational reform movement of the second half of the eighteenth century, this educational ideal was predominantly aimed at the middling orders of society, the wealthy German *Bürger* or bourgeois.

Book and box, theologian and salesman, child and Bürger are the keys to this study, the first on the subject, of Stoy's Picture Academy. The accounts of this extravagant and laborious enterprise offered by the existing secondary literature on children's books and on the history of pedagogy are limited in scope. The reasons for this may lie in the scarcity of sources concerning the author and his work, or simply in Stoy's obscurity.[2] Unlike the far better known Johann Bernhard Basedow and his book, the *Elementarwerk,* Stoy is not noteworthy for groundbreaking theoretical innovations in the educational field, nor for his skills as a merchant.[3] Thus, apart from some brief articles and isolated entries in reference works, the handbook on children's literature by Theodor Brüggemann and Hans-Heino Ewers offers the most thorough and complete description of the work and its author.[4] From the art-historical perspective, Brigitte Thanner deals with a portion of the copper engravings in a study of the Swiss artist Johann Rudolf Schellenberg, and Christiane Reuter has identified the sources for the pictures in the Academy.[5]

Drawing upon these writings, and upon previously neglected sources, this book offers a detailed presentation of the Picture Academy for the Young, its development, and its structure. It should be said at once that neither the Academy nor Stoy himself had any appreciable effect upon either the edu-

---

2. Stoy's papers seem not to have been preserved. Letters, certificates, and even cursory file entries are few and far between (see Unpublished Sources in the bibliography).

3. See Fritzsch's introduction to Basedow's illustrated primer *Elementarwerk* (1909) and Stach 1974. This reformer and his work, which is probably the most famous German teaching aid for children, is discussed here in chapter 2 in the section "The Eighteenth-Century Children's Book."

4. This work also provides the first complete bibliographic summary with information about contemporary and subsequent receptions of Stoy's work (see Brüggemann and Ewers 1982–91, 1099–114, 1536–38). Beyond that, an unpublished essay from 1981 cites a few textual and illustration sources and covers similarities with, and differences from, Basedow's *Elementarwerk* (see Reich 1981).

5. In her dissertation on this artist and engraver, Thanner (1987) investigates his commissions from Stoy. Reuter (1994) completely identifies the original illustration sources for the Academy.

cational doctrines or the forms of artistic representation current in the late eighteenth and early nineteenth centuries. What then makes this man and his project so interesting to us? Or put otherwise: What is so fascinating about this picture box to warrant a book-length treatment? Stoy is an epitome of the Enlightenment. All his activities and his thoughts about God, pictures, and education are the very essence of how Enlightenment was understood toward the end of the eighteenth century. And the box itself is an epitome of the Enlightenment. This single object shows the material basis for the "century of order"; putting things into a box is not only a daily activity but also affects mental categories. This point is of great importance for a history of the presentation and communication of knowledge. I explore particularly the tensions or differences between the traditional media of knowledge, books, and pictures, and a medium of presenting knowledge that does not originate in the field of education, the box. My aim is to pursue the implications of the assumption that an object can be treated as a historical source.

As early as the beginning of the eighteenth century, the Pietist and founder of an orphanage August Hermann Francke in Halle began to translate the world and its realia into a didactic medium.[6] He did so by encouraging the building of models and the collection of natural objects, which were then presented to children. In Stoy's case we are concerned with copperplate engravings. In them are images from nature, mythology, and biblical history, which express the significance of the Divine Creator. The representations from different domains of knowledge are always related to a biblical center, however, given meaning within biblical history. With Francke, natural objects could be ordered within the kingdom of nature according to a classification; with Stoy, biblical history was the means of unifying his disparate themes. In both cases, natural objects and picture cards functioned as individual, material carriers of knowledge.

The box, on the other hand, ordered the pictures and conferred upon them a comprehensive unity and identity. As Stoy asserted, "There may be nothing in the world that cannot be placed under these 468 rubrics" (1780–84, directions, 9). Clearly, it was the intent, not that the whole world would fit into the box, but that the box's order would inspire further additions of pictures and texts (quotations and definitions, for example) to the existing compartments. While the pictures refer to a linking theme, that of biblical history, which is also expressed upon the uncut tableau or picture plate, the box itself suggests that knowledge is not so much ordered by, and understood through, common

6. For the concept of realia and "realist education," see chapter 2, the section "Nonfiction Books."

meanings, but derives from a process of accumulation that requires constant revision.

From this arises a question, which directly derives from the medium: What is the difference between the pages of a book, which have to be turned, and the cards of a box, which must be taken out? By considering how the picture encyclopedia turns into a box, we can clearly see that a history of the perception of pictures does *not* merely concern the sense of sight, as has so often been suggested for the eighteenth century.[7] It must also take into account those modes of apprehension that involve the sense of touch, the haptic sense. The pictures stored in the box show that the development of the child's sense of touch played a particularly important role in the eighteenth century. This educational device, this toy, manifests a relationship between hands and words, as well as between hands and the objects shown on the pictures. In this sense, the box presents us with the act of taking possession of the world, the subject's ability to reach for something "always to hand" (Heidegger), which can appear as a whole only when contained within a box.

It was my personal fascination with this object that was the initial impulse in choosing what to treat historically. A picture box from the eighteenth century that is constructed like a modern-day card catalog: how can such an anachronistic object fit in the context of the Enlightenment? This wonder led to a cultural history, replete with surprises (Darnton 1984, 5), in which the Picture Academy is embedded. History—in the broadest sense—of mentalities and of everyday life served not only as a starting point but also as a stimulus for this book.[8] The gentrification of men and women of the lower classes, access to unusual, apparently out-of-the-way sources, and a selection of equally unusual issues—if not expressed and only sparingly cited—undergird this book. Darnton's analysis (1979) of Diderot and d'Alembert's *Encyclopédie* with its economics and publishing technicalities provided impetus for a biblio-historical glance at Stoy's work. During the course of my research it became apparent that the structural arrangement of the Picture Academy and its medial transposition into a box of pictures are important. From this perspective, a number of eighteenth-century boxes of traders, collectors, scientists, and educators were thrown open that had hitherto gone unnoticed by historians. The idea of the box, which had seemed unique, was revealed as something quite common in

7. For an in-depth study of this aspect, see Jay 1988, which discusses the various ocular models of the early modern period and investigates "the plurality of scopic regimes" (p. 20).

8. History of mentalities and of everyday life, as is appreciated, e.g., in Davis 1983, Ginzburg 1983, Honegger 1977, Raulff 1987, and Sabean 1984.

its day. The central question now is, Why was a pictorial encyclopedia of the Enlightenment transferred into a box?

The Age of Enlightenment was also an age of order. Michel Foucault's description (1966) of the structuring principle or the connection between *les mots et les choses* and his analysis of the orders of the classical era in the fields of natural history, economy, and linguistics showed that even widely separated thematic fields may be related to one another. On what basis can disparate things be related? Foucault considers the layers underlying such ordering as *epistemes* of Western thought. Building upon this interconnection between intrinsically heterogeneous subjects, the following history of an encyclopedic picture book proposes that, when examining the *epistemes,* material culture must not be neglected. Could Linnaeus have developed his classification without his compartmentalized herbarium cupboard, without the meticulously measured beds of the botanical gardens in Uppsala? Although these questions are left unanswered here, they do sharpen the focus of investigation, namely, the connection between verbalized versus visualized theory and actual practice in the eighteenth century.

To establish a link between theory and practice, between book, image, and box, requires a transdisciplinary approach and disparate sources.[9] As aids, I used recent historiographic models in history of art, collecting, and the sciences. The function of an image within its historical context and the vehicles available to it are issues raised in a recent branch of art history, according to which an image relays content and comprehends meaning that are not based solely on the artist's intent and are determined by more than just an inherited formal language.[10] In this examination of the Picture Academy, prominence is given less to the relationship between the contractor of the engravings and his artists and printers, along with their conceptions of what an image entails,[11] than to the vehicles for the copper impressions and their arrangement. How is a depiction handled, ordered, and stored for a given purpose? What significance does it gain with regard to cultivating the eighteenth-century child's mind and perception? That the picture cards were supposed to be "grasped"—in both

---

9. References to the literature have been kept to a minimum in this introduction. More complete citations and comments on the relevant secondary sources can be found at the beginning of each chapter.

10. Stafford has recently produced a number of studies on the functionalization of images; for the eighteenth century, see Stafford 1991. Haskell (1993) discusses what pictures can say about the period in which they are made and their value as a source. See also Beyer (1992), who presents the volume he edits as a reflection upon the art historical method of iconology. Raff (1994) refers to an iconology of materials in art.

11. Emma Spary (1999, 358) legitimately seeks such a connection in her review of the German first edition of this book; the lack of source material, however, unfortunately precludes this.

senses of the word—supports the contention that a copperplate engraving or picture card should be regarded as a physical object and be associated with the history of physical things.[12] This material culture of a period, a subject that has increasingly penetrated the concerns of historians, has found its most important expression in the history of collecting. The study of material culture is not a discipline in the strict sense, like art history, but it unites a history of objects with the central historical phenomenon of the seventeenth and eighteenth centuries: collecting. Here we are dealing with not only the question "What objects did they have?" but also the question "What did they do with them?" (Herklotz 1994, 123).

This problem is of importance in the history of science as well, as is shown most prominently in recent papers on historical scientific experiments. The realization that the success of an experiment rests not merely on tools and apparatus, but just as much on bodily experience and the establishment of a personal experimental space, led to a redefinition of the relationship between theory and practice in the sciences.[13] Such analyses of scientific spaces and objects pay attention to science within the network of social and economic conditions.[14] Science moves away from the realm of fact and into correlations of "making" (Rheinberger) and its conditions. Although no specific scientific process can be linked to boxes, this summary of the historical perspective on science clarifies the "making" aspect of the box, the formative occupation with a thing as a man-made tool.[15] Furthermore, the box (as we have seen above), understood as a framework of learning and therefore as a specific representation of knowledge, lays the groundwork for the engendering of knowledge, which—certainly in the eighteenth century—cannot be clearly isolated from the emerging sciences. For boxes provided order in the eighteenth century. Denis Diderot, in connection with the ordering of knowledge, spoke of a *mappemonde,* a world map of knowledge, which he circumscribed

12. Such a history drew its inspiration above all from the history of mentalities, in which importance is attached to simple everyday objects (see Raulff 1987, especially the contribution by Le Goff, 18–32, as well as the key work by Braudel 1981). In the field of sociology see Baudrillard 1968 and Bourdieu 1977. Within the field of German folklore studies *(Volkskunde),* this subdiscipline is stabilizing nowadays under the heading *Sachkulturforschung,* or the study of objects in terms of their sociocultural function (see Hauser 1994, 12). Within art history, a classic work is by Siegfried Giedion (1948); and George Kubler (1963) discusses the "shape of time" as a history of things. On the intersection points between histories of art and science in this field, see Rheinberger 1992, 17. For an anthropological perspective see Miller 1998.

13. See Sibum 1995; on experimental spaces see in particular Shapin 1988, Ophir and Shapin 1991, Outram 1995; see also Rheinberger et al. 1997.

14. Cf. Bijker et al. 1987, Gooding et al. 1989, Latour 1987, and Shapin and Schaffer 1985.

15. Cf. Latour 1990.

at the beginning of his *Encyclopédie,* to serve as a guide for the user during his wanderings through the foreign fields of theory and fact; Stoy, in contrast, envisioned a boxed-in world, a *Weltkasten,* and his idea was to promote, not idle travels in the study, but diligent pulling and refiling of the elements of knowledge. In following Stoy's gripping enterprise, the object itself determines which historiographic approach this analysis must take. Piece by piece, compartment by compartment, my study advances toward an encyclopedic vision of the eighteenth century; and if I were to choose a term for this object-oriented investigation, it might be writing *tactile history.*

In this vein, we begin with the production and sale of the Picture Academy, as well as a biography of its author. The book itself is at the forefront: the first part describes and analyzes it in the context of a comprehensive study of the secondary literature. In the second part, I explore the sources for the book and its illustrations. I study the pictures of the Academy in relation to the representation of the German *Bürger,* with reference to contemporary models of the body. Finally, the third part deals with the material construction of the Picture Academy and follows the form of the box in the metaphors and media of the second half of the eighteenth century.

BOOK

1

THE WORK AND ITS AUTHOR

The Production and Reception of Stoy's Picture Academy for the Young

Even to the modern observer, when first leafing through the Picture Academy of the parson and pedagogue Johann Siegmund Stoy, it is obvious that the author proposed to present a whole, a picture collection, a kind of comprehensive storehouse. The pictures and the accompanying text are rich and detailed, so that one's first impression is of an unrelated chaos of pictures. But the work's title offers an order for the diverse fields of knowledge in the Academy, and thereby a means of orientation.

> Picture Academy for the Young. Illustration and Description of the "Most Notable Objects" of Youthful Attention—from Biblical and Profane History, from Ordinary Life, the Kingdom of Nature, and the Vocational Trades, from Pagan and Ancient Lore, from the Best Collections of Good Fables and Moral Tales—Besides an Excerpt from Mr. Basedow's *Elementarwerk.*

The most notable objects, or those of general knowledge, were collected in a "textbook and reader," and the complex contents indicated by the title were categorized according to a division of subjects established by Stoy. The facts deemed most important or worthwhile for children and young people to know were selected from the nine subject areas or disciplines and presented in the Academy as an instructive but also entertaining children's book.

The Academy consists of two volumes of explanatory text and a single volume with fifty-two plates of illustrations. While the text volumes are always bound in the same way, the picture tableaux are presented in different forms. The fifty-two plates of the Academy are each subdivided into nine fields (figures 2 and 3). Grouped around a large central picture are eight smaller ones, which are numbered in the same sequence for all fifty-two copperplates. Starting with the central field, the numbering is from left to right, as in reading. Each number, or field, has been assigned one of the main subjects mentioned

Figure 2. Copperplate 4 of the Picture Academy (Stoy 1780–84)

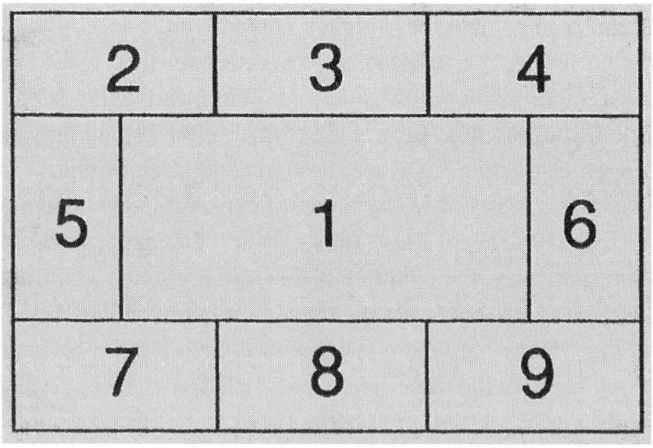

2	3	4
5	1	6
7	8	9

Figure 3. Layout of the fields in a tableau (diagram by the author)

in the work's title; these copperplates are preceded by two more full-page engravings (the frontispiece and the dedication plate).

"As the main foundation of the whole work I have chosen, so as not to embark on a completely new, usually misleading path, biblical Scripture, because it is the first way with which one commonly tends to engage the young" (Stoy 1780–84, foreword, 10). Of the fifty-two episodes from the Bible, two-thirds stem from the Old Testament and one-third from the New. This topical field was given a central position not only in the composition of each plate but also substantially, in that it provides the basic theme for all the smaller fields. "To biblical Scripture I have linked, mostly by a very natural connection, all that there is to tell the young, primarily in their first twelve years, about all areas of the sciences and about the most prominent appearances, occupations, and good and bad actions of mankind" (p. 11). The second field is devoted to ordinary life, although—contrary to what the title would lead us to expect—this does not refer to everyday life or general advice. This field rather serves as an omnium-gatherum of anecdotes, curiosities, and stories aimed at instilling a devout fear of God; its purpose is to impart general truths by means of allegory (see Stoy 1779, 5).

Field 3 is grounded in secular history and covers topics like travels to the New World; but biblical themes, such as the wanderings of the Israelites, are also associated with this field. Field 4 contains extracts from a similarly comprehensive children's book of the eighteenth century mentioned in the title, the *Elementarwerk* by Johann Bernhard Basedow, first published in 1774. Almost all the copperplate engravings of this work were included in the Picture Academy. The fifth field is dedicated to the kingdom of nature. From general representations of the animal and plant world, to human maladies and physical instruments, these pictures reflect everything that is included in the eighteenth-century word *Nature.* In the sixth field, Stoy grouped various trades, and for the seventh, he commissioned illustrations of Aesop's fables. Mythological depictions in the eighth field and moral instruction in the ninth completed the series of themes.

> The essence of this elementary work, however, which sets it apart from all the others, is the connection and relatedness of all the images of each plate with a main idea, which commonly is the biblical story upon which each plate is grounded. This ranking, ordering, and linking of the subjects is the soul of the work, and use of the same yields untold benefit. (Stoy 1791, 6)

Each copperplate has a central, biblically oriented premise upon which the other fields rest. The fourth copperplate, whose center depicts fratricide perpetrated by Cain (field 1), may serve here as an example (see figure 2): Cain strikes down his brother Abel with a stone. Stoy explained that this murder, and with it theft, anger, and wrath, were outcomes of the original sin. Nevertheless, people should not act as Cain did, but rather according to the maxim "Do no evil, and no evil shall be done unto you!" (Stoy 1780–84, connection between the images, iii). The allegorical field (no. 2) symbolizes this motto in a negative sense. Because the wolf tried to seize the lamb, he was killed by a hunter and in that way received due punishment. The third field, for ancient and secular history, shows the appropriate counterpart to the biblical episode, namely, the fratricide committed by Romulus against Remus. In the fourth field, the extract from the *Elementarwerk,* three situations are represented in which someone is trying to escape from mortal danger. One man is fleeing from an assailant who is brandishing a flail, a second is seeking refuge on the steps of a building from an ox, and the third is hanging onto a floating barrel to keep from drowning. In the fifth field, a crocodile, the "foremost butcher of men" (p. iii), represents the link between theft and murder. The illustration in the sixth field shows a surgeon wrenching a shoulder back into place. In the seventh field, the wolf again threatens the lamb, and in the eighth field, that of mythology, the Fates are spinning the thread of life, only to cut it. The ninth field, that of moral tales, teaches how marital anger and wrath are best avoided.

Thus the central theme of this tableau is murder or the intentional infliction of bodily harm upon a person. This underlying theme—personified by the biblical figures of Cain and Abel—Stoy employed to spin a web of thematic combinations and links between crocodiles, wolves, and doctors, in order to help children retain what they placed in their memories and use it in practice. To this end, individual episodes of a story—such as that of Romulus and Remus—must be singled out and depicted in necessarily abbreviated form. The communication difficulties arising from such a fermentation of knowledge will be discussed below (chapter 3). Besides the abbreviation, the noticeable lack of connection between the contents of some fields and the center also hampers the reading and understanding of the picture plates. A typical example of this is the ninth field. It shows an older woman behind a table, handing a bottle to a much younger one. The young woman has asked her elder how to keep her husband from beating her. With the bottle are exchanged rules and worldly wisdom that help prevent rage. The connection to the central theme cannot be made out from the illustration, taken by itself. Although there are two separate volumes of accompanying text, they do not always explain the multiple levels of meaning in the plates and Stoy's thematic conception of them. Here we are, already encountering some of the criticisms raised in contemporary reviews of which Stoy was in some cases aware.

The Academy appeared in installments over four years and was available by subscription. By 1784, the reader could have the complete work, with comprehensive commentary, in hand. With each delivery, a fascicle of copperplate engravings consisting of six printed sheets arrived in an envelope bearing the title and in some cases additional information.[1] In the first two years of their appearance, the plates were merely accompanied by a text entitled "Connection between the Images." These brief explications presented each tableau within its thematic interlacement. Generally speaking, the purchaser was ultimately responsible for binding the whole Academy, as was true for most books of this period. "This storehouse can be made still richer," suggested Stoy, "by interleaving the copperplates with heavy sheets of paper, onto which one can glue pictures or drawings appropriate to each plate as time goes by" (1780–84, directions, 3). Stoy recommended the addition of blank pages to the bound picture book to allow the child to add pertinent pictures or texts that he or she had discovered. The individual plates could also be glued onto cards and hung on the wall. The teacher "has the plates mounted on thick cardboard and hangs them up in the room, either one after another or by and by all of them, if he has space" (p. 16). Once he had collected all the installments, the buyer could choose among these alternatives the exact form and binding of his Academy.

Stoy gave detailed instructions for the binding of the text. "The text refers to these copperplates and explains each tableau by its nine fields or numbers" (foreword, 12). Since the individual copperplate installments always appeared with a few explanations of their own as well, the explanations were supposed to be placed at the front of the first complete text volume. Stoy's advice to the bookbinders described the sequence in which the textual part of the work should be bound into a whole: "In the first volume belong [the following:] 1. the title page of the same, 2. the dedication, 3. the foreword, 4. the directions for the practical use of the Picture Academy . . . , 5. the sample of the different combinations of all the images of a plate . . . , 6. the connection between the

1. These envelopes survive only in a few cases (see the section "Binding and Jacket," in the appendix). An anonymous reviewer in the *Nürnbergische gelehrte Zeitung* of 15 November 1782 acknowledged their value, reporting that, unfortunately, "such blue envelopes are too easily lost" and what Stoy communicated on them could not be simply passed over "without a word" (p. 736). This instance related to some problems that Stoy encountered regarding his critics and the sale of his book. It can be assumed that such envelopes also bore information about the printers or sales outlets.

images. . . . 7. the explanation itself" (ibid., to the bookbinders). In the dedication, Stoy offered his work to the Swedish crown prince. In the foreword he discussed the meaning and purpose of the Academy, and in the directions he provided indications for its use. In the "sample," one plate served as a model for working out the possible sets of relations between the individual picture fields. The introductory material of the text volume ended with the "connections between the images," in other words, between the individual plates. This was followed by the "explanations," accounts or samples of stories concerning each field in the succession of the plates. Stoy supplied a text with its particular heading for each of the 468 individual pictures (fifty-two copperplates of nine fields each).

The second volume is structured in the same way. It likewise opens with the "connections between the images," followed by a commentary. It closes with a table of contents and a subject index. The table of contents lists all the images within a single subject field for each plate, from one to fifty-two, in chronological order. The index, which listed the "most curious things and persons" in alphabetical order, enabled the plates to be used independently of chronology, as a sort of pictorial dictionary or "as an encyclopedia . . . in which further [information] about this or that subject can be sought or looked up, when the opportunity presents itself" (directions, 13).

The Use and Handling of the Academy

> Although this work is being published in installments, and its actual practical use is to be determined only when it is complete, it will gain much recognition, recommendation, and further dissemination if I provide a brief sketch of it in the meantime, especially since I have been thus publicly urged to do so. (Stoy 1780–84)

In his introduction to the Academy, Stoy suggests three uses for it. First, he recommended it as a manual or as a "store of good food for the minds and hearts of the young." Such a manual was of particular value when children were not going to school or getting private tutoring. For adults, it could serve as a reference work and so substitute for an entire library: "the owner of the Picture Academy possesses not just one, but nine picture books, from the nine fields on each plate" (p. 2).

Both in private instruction and at school, the Academy could be used in the following manner. Every week one of the pictorial plates could be discussed with the older children (above ten years of age). It could serve as a suitable and comprehensive textbook by review of its content, consultation of other reference works, and study of additional illustrative material. Stoy had prepared for such a weekly learning program by providing fifty-two copperplates, corresponding to the fifty-two weeks of the year, so that young people could "grasp it in a year," as Stoy remarked in his prospectus for the Academy (1779, 3). Younger children of five to ten years of age could be taught exclusively with the Picture Academy. Initially, the teacher was not supposed to discuss the plates, so that the children could form their own impressions. Then, gradually, information about the depicted subjects was to be supplied. The teacher was not supposed to lecture from the text volume but to speak freely. "He takes the book into his own care so that the *élève* should not skim through the book too hastily and lose curiosity" (Stoy 1780–84, directions, 16). The children were to look at the book themselves, so that the knowledge they had experienced and learned once could be stabilized. These pictures could also be used as a reward: "In general, it would be good if every school had this or a similar picture collection in stock and always immediately at hand as a reward for the diligent" (p. 13).

In addition, Stoy stressed how important the field for "Vocational Trades" was for the older boys, since they could get a better conception of the future careers they might choose.[2] He mentioned the youths "who are later destined for the military, a trade, or craft and hence might dispense with academic presentation" (Stoy 1779, 30). Stoy explicitly remarked in his notice of 1782 that the Academy should be available at a price that even a *Mittelsmann* (p. 1), that is, a person from the middle classes, could afford to pay. This included public officials, such as secretaries and mayors, as well as merchants and tradesmen. Stoy also included within his target group a whole range of clerics (ministers, chaplains, preachers), along with academically educated teachers and rectors, as well as apothecaries, medical doctors, and jurists.[3]

Stoy recommended two ways for teachers to use the Academy. The pictures

2. As a rule, the term *children* was used to refer to boys or youths (young men). Just as the intended audience was largely composed of a male public, so the knowledge content offered in the Picture Academy predicated male preceptors to teach them (cf. the illustrations in this volume, especially plates 2, 4, 38, 44, 45). Although this gender-specific orientation is not taken up again here, let me allude to the reviews of the Picture Academy, e.g., in *Allgemeines Verzeichnis neuer Bücher* 1779, vol. 4, in which the reviewer speaks exclusively of young boys. Wild (1990) characterizes children's literature from the Enlightenment as "profoundly patriarchal" (p. 69); see also Wild 1987. On the education of children by their mothers and on the education of girls, see Kersting 1992, 275–389, and Kleinau and Opitz 1996.

3. For a more detailed definition of the German middle class of the eighteenth century, see Haltern 1985, 82–89, who locates the expressions *middle class, middling condition,* and *petit bourgeoisie (Mittelklasse, Mittelstand,* and *Kleinbürgertum)* within various models of social stratification. As in this instance, Stoy mostly used the term *Mittelsmann,* by which he meant a member of the middle class and not, as dictionaries of that time indicate, someone who mediates between two persons.

could be presented piecemeal. The emphasis here was on avoiding overloading a child's mind with too many pictures, and on simplifying the confusing multiplicity of combinations. To that end, Stoy (1779, 29) proposed that one begin with biblical history and present the child specific sections of the individual plates. Particularly in the first years of learning, it was important for the teacher to leave out the secondary associations (fields 2 to 9), "or he should completely cover them with a sheet of paper that has a hole cut in the middle, so that the child is not distracted by the many images." Another selective presentation method was mounting the individual copperplates on cardboard for hanging up in the classroom (see Stoy 1780–84, directions, 16). This method would ensure that the children did not page ahead of the lesson, become "dulled" by the many impressions, and lose interest.

The second method was collection of knowledge by the children themselves. To that end, the teacher could have the copperplates interleaved with sheets of paper (p. 3). On these empty pages between the plates, the pupils could gradually enter additional knowledge not contained in the Academy. The pupil could treat the printed text in the same way, so that "there may be nothing in the world that cannot be placed under these 468 rubrics, which can be still further subdivided" (p. 9). Stoy drew justification for this procedure from domestic economy: "The orderly housekeeper puts everything away in its proper place, whence he can find and use it again as he will—why should this not also be possible with the gems of learning, the sciences, and experience?" (p. 10). The underlying maxim is the attempt to give order to knowledge. Two types of order were involved here. The first was for the benefit of the teacher or the parents. In Stoy's opinion it was difficult to orient oneself in the deluge of available educational literature. For this reason his Academy incorporated a selection of the best writings and hence absolved the educator of any decisions about suitable reading material. The other type of order that the Academy aimed to impose concerned child education. Here, personal practice would ensure that order became an ingrained habit—a virtue that Stoy, like others in the eighteenth century, held to be of incalculable value for children.

The picture cards, stored in a specially constructed wooden box, made possible the selective presentation of pictures. The plates were cut up by subject, and each card was strengthened with cardboard, so that it could be shown individually. The subject divisions in the box corresponded to those on the picture plates, and all the cards of a particular field were filed in the same compartment (see figure 4). This presentational format, according to Stoy, was particularly suited for comparing neighboring images within a single category.

Figure 4. Picture box; the compartments correspond to the tableau divisions (photograph by the author)

When playing with the cards, a child could bolster his memory by the three-dimensionality of the picture cards.

The order of this human-oriented world was thus conveyed via word and image. The engravings and accompanying texts together formed a complementary whole. For younger pupils, the focus was on the plates; for the older ones, it was on the addition of new pieces of knowledge to the Academy's scheme. The textbook was intended for the teacher, as another author of children's books pointed out: "Books of this sort serve more as a guideline for the teacher, rather than having to reach the limit of the learner's knowledge" (Adelung 1771, xiv). The text did not always deploy stimulating and balanced language and could only brush over many topics; the teacher's role was to elaborate them further. Stoy viewed his method of ordering knowledge and pictures as the ideal condition for the learning child. Knowledge should be presented to him in "playful instruction" (Stoy 1782, 1); knowledge acquired through play was easier to retain (see 1780–84, directions, 11). Once children had appropriated knowledge for themselves, it only remained for that knowledge to be fixed in their minds through constant repetition, and thus to be definitively stored there.

The Production of the Academy

"Nothing has been spared for the advancement of the external quality and beauty; and warmest thanks must be extended to Mr. Chodowiecky and likewise to Mr. Schellenberg and Mr. Penzel for supplying and finishing the copperplates" (Stoy 1780–84, foreword, 13). The engravings, some colored subsequently in Nuremberg, had been prepared by the most famous engravers and draftsmen of the time. The fact that Stoy was able to acquire Daniel N. Chodowiecki and Johann Rudolf Schellenberg for his project reveals something of his acumen in business matters; and the Academy received much praise, not least for its aesthetic qualities. These two artists also coordinated the work to be done. In 1780 Stoy thanked Chodowiecki "both for most generously assuming the supervision of this work right at the outset and for the devotion, pains, and care he has lavished on it thus far."[4]

The other draftsmen engaged to work on the Academy were Daniel N. Chodowiecki's brother Gottfried and Johann Georg Penzel. Besides Schellenberg and Penzel, other engravers' signatures included Johann Friedrich Schleuen, Carl Christian Glaßbach, F. Conrad Krüger, and Johann Carl Bock.[5] Stoy mainly corresponded with Chodowiecki and Schellenberg, however, and he negotiated the costs and settled accounts with them. For example, he complained to Chodowiecki that the engraver Glaßbach was "taking far too long about it," and promised Chodowiecki's "dear brother as the engraver . . . 1 extra louis d'or" in exchange for which, "however, we respectfully beg for more detail and accuracy."[6]

The surviving correspondence between Stoy and Chodowiecki gives an impression of how Stoy communicated his conceptions of the pictures to his draftsmen and engravers. First he sent a list of the individual fields of a plate and a short account of their content to the artists. In making the drawing, the artists usually relied on existing illustrations from other textbooks to prepare initial sketches, which were sent to Stoy. It may be assumed that Stoy had suggested at least some of the books from which the illustrations were selected, and Daniel N. Chodowiecki, who by this time was an established illustrator of children's and juvenile literature, could tap into his store of work completed for earlier publications (such as Basedow's *Elementarwerk).*[7] After Stoy had

4. Undated letter, Darmstaedter Collection 2d1778(6), Staatsbibliothek zu Berlin Preußischer Kulturbesitz.

5. The engraver and artist could not be definitely ascertained for every plate; see the appendix.

6. Letter dated 19 March 1780, Darmstaedter Collection 2d1778(6), Staatsbibliothek zu Berlin Preußischer Kulturbesitz.

7. In her thesis on Chodowiecki and eighteenth-century pedagogy, Maria Ledderhose (1982)

examined the drawings, he sent them back with some requests for correction. Then the copperplate was engraved in Berlin or in Winterthur and printed by the publishers Weigel & Schneider, later by Stoy's private press. Having the plates printed in Nuremberg was not necessarily the normal procedure. So, for example, Daniel N. Chodowiecki's contract for the production of the engravings for the *Elementarwerk* required him to supply actual prints, not just the copperplate with the engraving (see Gilow in Basedow 1909, 3:4 f.). In the case of Stoy's Academy, however, the prints from the first six installments bear the watermark of the Weigel publishing house (see the appendix). Schellenberg's bill, too, includes the cost only of drawings and etched plates, but not of prints (cf. Thanner 1987, 365–67).[8]

The two sheets originally designed for the Academy were the frontispiece and the dedication plate. The title page was drawn by Daniel N. Chodowiecki and engraved by Schellenberg. Earlier, on 6 September 1782, Stoy had sent the draft sketch back to Chodowiecki in Berlin with a few remarks and corrections. He requested numerous detailed alterations, which reveal that he not only had strong opinions about the design, but by that time was able to define the focus of the work: "So that all nine divisions of this elementary work can be shown figuratively on this page, the wall that Clio is unveiling is dedicated to biblical and secular history. . . . In this way the parallel between spiritual and secular history, or the main purpose of the book, becomes clear."[9] The correspondence about the frontispiece continued for some time, and the frontispiece was at last included in the ninth and final installment.

Because Stoy drew text extensively from the same sources from which he had borrowed the illustrations (which was then common practice), we cannot rightly refer to him as the author of the Academy. He is more appropriately described as its conceiver or eclectic editor, even though in places the text was originally his. He reported in his foreword:

> I have scoured the whole great stock of educational works that have appeared in the last 15 to 20 years, during which period I was preparing the Picture

---

analyzes the Berlin artist specifically as an illustrator of pedagogical works. This thesis is a useful survey, since it lists Chodowiecki's work and identifies its characteristic attributes. Ledderhose does not, however, embed her study in any deeper historical context. On that question see the article by Schmitt on Chodowiecki as a Philanthropinist illustrator (in Schmitt et al. 1997). A more general classification of the work of this artist as an illustrator is a volume on Chodowiecki and the Enlightenment, Rothe and Ryszkiewicz 1986.

8. On Schellenberg see Thanner 1987 and the volume concerning his role as an illustrator of natural history works (Thanner et al. 1987).

9. Letter of 6 September 1782, Darmstaedter Collection 2d1778(6), Staatsbibliothek zu Berlin Preußischer Kulturbesitz.

> Academy—and did not shrink from the labor of using the best and most comprehensive works for certain divisions, particularly in the kingdom of nature and the vocational trades, and collecting the best material for primary instruction from them. (1780–84, foreword, 12)

The Academy was, consequently, an anthology of passages and images from various extant works. This had the advantage that such an encyclopedic work for children, a comprehensive project, could be completed swiftly and thereby reach the hands of subscribers at reassuringly short intervals. Stoy had to reconcile himself to not being acknowledged as its true author, however, and it was precisely this point that Stoy's contemporary critics took up.

Contemporary Reviews

The composition of the Academy led Stoy to be charged with "blindly producing an epitome" (Stoy 1780–84, 12). This was a common criticism in the eighteenth century, particularly of academic works thought to have been written purely for financial gain. Stoy, among many other writers, was held to account for a literary form that at the beginning of the century had still been considered legitimate: analyses treating the validity and purpose of established, published knowledge.[10]

A contemporary review of the Picture Academy reads: "Although the book may be put to profitable use in public and private education, since it is, in the main, a mere compilation, surely we should have expected something better" (Baur 1790, 483). Even considering that in the eighteenth century "texts for children were considered a form of public property" which was "available to anyone wanting to dabble in children's literature" (Wild 1990, 75), and that reuse of existing sources was not deemed exceptional, this critical formulation was biting nonetheless; and it was immediately followed by the comment that "the Bible, the songbook, and the other gamebooks for the age of childhood, which Mr. Stoy has published, are of about the same value as the Picture Academy" (Baur 1790, 483). Stoy evidently anticipated this reproof even before the complete Picture Academy had come out, since he defended himself against it in the foreword to the first text volume: "Upon close inspection of

10. "Another favorite means of commercial book production was the making of new books out of old. With the spirit of translation came the prevailing spirit of compilation, arising out of the same causes, characterized by the same qualities in the public, the writer, the publisher; the spirit of writing up, writing out, writing after, in whose lowest forms plagiarism and reprinting can often scarcely be distinguished" (Goldfriedrich 1909, 307). Cf. on "epitomes . . . and other varieties of impropriety" Johns 1998, 30.

the text, [the charge] of blindly producing a compilation or epitome will fall on its own" (Stoy 1780–84, foreword, 12).

The Picture Academy is certainly no "blind" compilation. Nevertheless, it is quickly obvious that virtually all of Stoy's publications are revisions of texts and illustrations from other sources. In a review of his "Golden Mirror" (*Der goldene Spiegel* [1778–81], one of Stoy's children's books predating his Academy), this fact was pointedly referred to: "The perpetual, aimless gleaning from children's books for children's books has now become so much the fashion that the very sight of each new product of this kind almost moves one to disgust" (*Nürnbergische gelehrte Zeitung,* 28 May 1779, 103). But thanks to the fine copperplates accompanying each story in Vicar Stoy's publication, the reviewer continued, this reaction was made considerably milder, which verdict was shared by other reviewers as well.

Indeed, at the time a carefully illustrated work was still out of the ordinary, so the engravings helped to soften harsh attacks. "This Picture Academy snatches away from children many a painted and engraved piece of nonsense, whose form and material spoil both mind and taste!" (*Nürnbergische gelehrte Zeitung,* 22 October 1779, 695). "The copperplates . . . depict the subjects very vividly and naturally" (*Allgemeines Verzeichnis neuer Bücher* 1782, 7:380). In this context, the reviewers also specifically mentioned the high-quality paper and excellent prints, which indicates that a physical value was attached to the book. Just as Robert Darnton (1979, 179–80) has shown to be the case for French readers in the prerevolutionary period, who carefully examined a book and the quality of its materials before buying, the critics and purchasers of the Academy attended to the skill with which it was produced. So great was the interest in the book's physical trappings that the improvements in quality introduced with the second installment of the Academy (*Nürnbergische gelehrte Zeitung,* 13 October 1780, 662) could be traced back to Stoy's leaving his clerical career and his new exclusive focus on the production of the Picture Academy (*Allgemeine Bibliothek* 1782, 10:463).

Notwithstanding the general praise for the copperplates, reservations were expressed about the size of some of the images: "The pictures often err by being too small, which makes using them on dark days and by night inconvenient" (*Allgemeine Bibliothek* 1780, 8:388). This criticism, which was justified on didactic grounds, was directly bound to a further shortcoming: "With such a swamping of images, the boy becomes bewildered, and bewilderment leads to distaste" (*Allgemeines Verzeichnis neuer Bücher* 1779, 4:613). The complaint that Stoy had designed plates so small that they were difficult to use for teaching was one of the central points of criticism of the Academy. Even the text could lead to confusion, since the author did not content himself with a description of the pictures on each plate, but instead "he immediately divides all the plants, animals, arts, and sciences, into all their classes and subclasses; but only gives an outline of them and does not consider that, while it is easy for him to represent all these divisions and subdivisions in skeletal form from some book or other, not a single boy benefits from seeing all these skeletons" (ibid.). The description of a tea plant and its illustration, for instance, Stoy took as an opportunity to classify and introduce all the useful exotic plants. These were described neither verbally nor visually, however, but merely named. The reviewer exclaimed, "The teacher's work is aggravated here, and truly not simplified" (ibid.), and finally asked, "Is it not to be feared that the excessive struggles for some enlightenment will imprison the understanding?" (p. 614). The carefully conceived order of the Academy, which initially promised the opposite of confusion, collapsed under its own weight, according to the reviewer, when Stoy overburdened the relevant subjects and allowed a single plant in the picture to serve as proxy for an entire class. For these critics, the positive aspects of the Academy certainly did not outweigh this "imprisonment of the understanding."

Critical remarks were also directed at the arrangement of the individual fields around the center of biblical history. The correlations with the first field seemed "forced, at times" (*Allgemeine Bibliothek* 1780, 8:385), in other words, illogical and not understandable in context. Even when, for example, the criticism of Basedow's *Elementarwerk* was particularly strong over just this question of a lack of orderliness—"For Basedow's *Elementarwerk* is a very handsome collection of good and useful knowledge, but it cannot be denied that this collection lacks the necessary order, and the facts do not succeed one another the way they would in an improved lesson" (*Allgemeines Verzeichnis neuer Bücher* 1780, 5:219)—Stoy's reviewer could not "by any means approve of this method of regarding everything in relation to religion" (*Allgemeines Verzeichnis neuer Bücher* 1783, 8:685). The order of the images was an important criterion for the reviewers, but that chosen by Stoy was not adequately worked out, and only a "skillful teacher" could "know how to make this work useful for the youth of any rank, age, and origin" (*Allgemeines Verzeichnis neuer Bücher* 1782, 7:381). Despite more faultfinding—errors in content, overly complex vocabulary, the occasional colloquialism (*Allgemeine Bibliothek* 1780, 8:389), and "shallow declamation" (Baur 1790, 483)—Stoy was ultimately credited for having provided a secure foundation of knowledge for children with his work (*Allgemeine Bibliothek* 1780, 8:387).[11]

11. In *Kinderfreund* by Christian Felix Weiße (the most famous German children's periodical

Such reviews may be found both in professional educational journals and in general literary periodicals. The Academy was reviewed over pretty much the whole period during which the installments appeared, although with decreasing frequency toward the end of its publication.

Price Comparisons

As the reviewer for the periodical *Allgemeine Bibliothek für das Schul- und Erziehungswesen* of 1782 pointed out, even "for the seventh explanatory issue, which has a larger number of printed sheets," the price had not gone up, "which truly does credit to the auth[or]'s altruistic zeal to serve his fellow men in all possible ways" (10:464). While the Picture Academy was still being printed, in 1782 Stoy noted that the engravings alone cost one carolin, the text volumes three florins and thirty-six kreuzer.[12] The total price for a purchaser who subscribed in advance or who commissioned a copy from Stoy was eight taler. A buyer who did not want to commit himself to a subscription, but preferred to pay in installments, would have to pay an extra taler (Stoy 1782, 4). The final price for the complete Academy in 1784 was two louis d'or, equivalent to ten taler; this price remained the same the following year. Consequently, it was cheaper to place an order for the Academy at the first delivery of the individual installments. To offer a publication project like the Academy for subscription was quite in keeping with the usual buying conditions.

A comparable case is the first volume of Johann Samuel Halle's children's encyclopedia, which with one copperplate engraving cost one taler in 1779. A successor to the *Orbis pictus,* the *Schauplatz der Natur und Kunst in 4 Sprachen,* a single volume containing many engravings, was priced at three taler in 1781. This shows that engravings were by far the most expensive element of such publications and that their number determined the cost of a work. Another children's book of similarly labor-intensive design was Basedow's *Elementarwerk.* The total price was twelve taler, the cost of the text alone four

---

of the Enlightenment period), Stoy's Picture Academy was recommended to children for the new year, along with philanthropic children's literature (see Hurrelmann 1982, 206, 218).

12. One louis d'or, or one carolin d'or = eleven florins, or five–six taler; one florin = sixty kreuzer = one gulden = about half a taler. In the literature, the exact currency values differ slightly; the following sources were consulted for this comparison: Busse 1795–96, 4; Fickert 1989, 39; Leder 1965, 68; Roth 1800–1802, 4:250–57; and Stoy 1782, 4. Generally the gulden was worth less than the taler. In the northern German states, the groschen and taler were more commonly used; in the south, the gulden and kreuzer. French currencies like the louis d'or were more prevalent in southern parts. Stoy had to indicate a conversion rate because, like the laubtaler or the engelstaler, the carolin or florin fluctuated in value. (The Rhenish currency was already established in the city of Nuremberg, whereas the surrounding areas continued to use the Frankish currency for some years.) See also Stoy 1780–84, 982 f.

taler, and of the engravings on their own eight taler. From time to time the second edition of a work was cheaper; the price of Johann Christoph Adelung's *Unterweisung in den vornehmsten Künsten und Wissenschaften, zum Nutzen der niedern Schulen* (an elementary book by the well-known German journal editor), containing forty-one plates on foldout inserts and 512 text pages, fell from thirty taler in 1771 to twenty-two taler in 1785. The cost of Stoy's work reflects the state of pricing revealed here. This comparison does indicate, however, that the lack of change in the price cannot be attributed solely to "the auth[or]'s altruistic zeal." On one hand, Stoy had to worry about a drop in sales if he allowed the price to rise because of a few additional sheets. On the other hand, he still hoped at this point that the Academy would continue to sell past the first printing and that the initial costs primarily incurred by the artwork would then be redeemed.

If one compares the cost of the complete Academy to the average annual salaries of the period, it is not difficult to see that the buyers were members of the upper middle class and nobility. A Nuremberg painter or *Buntmaler,* for example, earned 400 florins a year—as much as the annual budget of an Altdorf student—a privy councillor, by contrast, 1,050 florins. Consequently, many among the populace bought for their children cheap prints for a couple of kreuzer from one of the many vendors in Nuremberg—a literature Stoy referred to as "pathetic daub and drivel" (1789, 5). The *Mittelsmann,* to whom the Academy was addressed, could not easily come by a copy of the book; in this instance, the intended readership and the potential market did not coincide.

An obvious question is how Stoy, himself a *Mittelsmann,* had obtained the capital necessary for publishing his Academy. A legacy at the beginning of the 1780s provided the opportunity for him to strike out on his own in 1782. The last will and testament he drew up with his wife, Barbara Magdalena Stoy, in 1777 shows that at that time his funds were insufficient to finance the Academy.[13] Against the costs that Basedow incurred for his *Elementarwerk* (before printing began he had raised about fifteen thousand taler), Stoy's publication of the Academy, although still a daring and expensive enterprise, does not appear particularly remarkable. Even though Basedow's project of a few years earlier required substantially more collaborators on the various subjects, its costs in relation to hiring draftsmen and engravers for the individual plates are comparable. In 1773 Basedow paid thirty reichstaler for an etched plate, which included a contract specifying a run of two thousand prints. That sum included the artists' salaries and the cost of the materials. The cost of

---

13. See the will of 21 April 1777, rep. 92, Reichsstadt Nürnberg, Testament, no. 1118, Staatsarchiv Nürnberg.

engraving a single plate ran to ten reichstaler (cf. Gilow in Basedow 1909, 3:12). In the itemization by Schellenberg, who worked for Stoy until 1789 (see Thanner 1987, 365–67), the following entries are under the heading concerning the plates: for one etched plate Schellenberg requested a sum ranging between thirty and fifty gulden, and for a drawing, a sum ranging between ten and fifteen gulden. In terms of price, the Picture Academy is not essentially different from similar books published in the last third of the eighteenth century. While it may never have approached the fame of Basedow's book, it sold well enough that various installments went through several printings.

Johann Siegmund Stoy

The armorial bearing of "Hr. Ioh. Sigmund Stoy Professor der Paedagogic" is displayed in a book of heraldry by the copperplate engraver Martin Tyroff (figure 5). This book, which appeared in 1783 in Nuremberg, depicts a collection of escutcheons of various distinguished *Bürger.* First the coats of arms of jurists are presented, followed by those of preachers, and finally professors. The clergy fill the subsequent plates, followed by pedagogues. The name Johann Siegmund Stoy[14] appears under the heading "Nuremberg Teachers and Schoolmasters"; his neighbors "in arms" are a school rector and a deputy headmaster by profession. Here we encounter Stoy's name with the gloss "Professor der Paedagogic," although the hierarchy in the book clearly says that he was a professor of lower rank.

Stoy's escutcheon is ornamented in the standard heraldic form, with a crested helm in the center and bushes of feathers mantling the actual shield. The helm crest is the upper body of a youth rising from between two trumpet horns. The youth holds a hammer in his right hand and a miniature of the helm in his left and appears to be working the helm, holding up the objects with his arms outstretched. Hammer and helm are therefore the guiding symbols for the name and person of Stoy. In the Academy parts appearing at the time, Stoy identified these symbols as the outer trappings of Vulcan, the god of fire and of all artisans "who handle fire and are involved with smelting, forging, and other types of metalworking; that is why he is also commonly represented as a blacksmith sitting at an anvil, holding a hammer in his right hand and a pair of tongs in his left, and forging a helmet for Pallas" (Stoy 1780–84, 133 f.).

Insignia, coats of arms, and orders of distinction are what Stoy described in this passage for the twelfth plate of the Picture Academy: "The essentials

14. The contemporary and current spellings of his middle name alternate among Siegmund, Sigmund, and Sigismund.

Figure 5. Armorial bearing of Johann Siegmund Stoy
(Tyroff 1783, plate 21)

of an escutcheon are the figure of the field or shield and its partitioning; the tincture or color; the charge or figure that appears in this shield or a differently fabricated one" (p. 191). In his own armorial bearing, the dancetté charge forms four interlocking triangles, the colors are ruby and silver, and above it rises the figure of the young smith.

This coat of arms, presumably commissioned from Tyroff the year before, may convey something of the pedagogue's self-image, about which the few remaining sources are scarcely able to give information. It documents an event in Stoy's biography that took place one year before the publication of Tyroff's heraldry collection. On Saturday, 23 November 1782, the Governor's Office and Prefecture of the imperial city of Nuremberg granted Stoy's petition to resign "his clerical office because of his weakened state of health," in which he was no longer able to perform "as his duties require."[15] So Pastor Stoy quit

15. Vormund- und Land-Pflege-Amt, entry in Verlässe des Innern Rats, rep. 60a, Reichsstadt Nürnberg, no. 4125, Staatsarchiv Nürnberg.

his ministry. In a historical journal for the city of Nuremberg a matter-of-fact entry reads: "When Mr. Johann Sigmund Stoy relinquished his ecclesiastical benefice in Henfenfeld and moved to the city, he received from the Very Rev. Superior the title of a professor of pedagogy" (Von hiesigen Professoren 1787, 263 f.). The Governor's Office or Stoy's church *Scholarchat* had thus granted his petition for a career change.

Stoy moved out of his parsonage in Henfenfeld near Nuremberg and to the city nearby. It was important for him to publicize his newfound status as professor of pedagogy as soon as possible, and his coat of arms was a helpful tool in establishing himself. It is not difficult to see the link between the figure of the youth in Stoy's armorial bearing and the familiar eighteenth-century metaphor for the educational process: just as the young tree must be trained upright and straight over time, so must the metal be worked before it ceases to be warm and soft. Both of these metaphors are part of the language with which the education and development of the child was described in this period.[16] Even if Stoy found his future profession to be "so shamefully disdained,"[17] this image of his armorial bearings served him as a bookplate. The problems motivating him to abandon his naturally inherited vocation for other work arose eleven years earlier, however. An event in 1771 interrupted his previously unspectacular professional training.

In the eighteenth century the office of a preacher or parson was a political state position. It was a public office, and for this reason its proper exercise required impeccable conduct. At the time, the imperial city of Nuremberg and the Nuremberg Territory maintained various supervisory offices and agencies to oversee the religious authorities. There was no lack of admonitions and disciplinary measures from above "that were meant to lend weight to the internal obligations of the church, in the interest of the state, and to investigate infractions against the public order. All the more important cases, and such they usually were, were decided by the city council itself" (Haniel 1962, 354). The council of the free imperial city of Nuremberg was composed of forty-two members, thirty-four of whom had to come from families that had the right to sit in the council, in other words, the city's patricians. Eight of the members were artisans. The case of the young preacher Johann Siegmund Stoy was one of the more important ones taken up by the council in 1771: on 12 March of

16. On the Promethean theme alluded to here, see the article by Bilstein (1992), who derives a collection of images and metaphors for the development of the human being from artisanal and artistic forms.

17. In a notice of 20 February 1781 he remarked that "money could not buy" the "faithful rearing and teaching" of a boy, yet it is still "as shamefully disdained as if it were nothing" (Darmstaedter Collection 2d1778(6), Staatsbibliothek zu Berlin Preußischer Kulturbesitz).

that year, Stoy was reprimanded for "personal neglect in not pronouncing the prayer formula prescribed by the church authorities for the current price rises and grain shortage,"[18] and his license to preach was temporarily suspended. Clergymen were expected to conform to custom and behave irreproachably in religious matters, and Stoy was apparently not meeting the standards. The circumstances explain why. The years 1770, 1771, and 1772 were remarkable for bad weather throughout Europe, with catastrophic harvests as a result. There was a basic grain shortage, prices rose immeasurably, the lower orders were deprived of the necessities of life, and a famine broke out, lasting several years.[19] The church assumed the task of calming the populace, obeying the city council's order in the winter of 1770 for the clergy to read "chapters of the Bible that are principally applicable, in these extremely troubling times" (Abel 1981, 30). A list of suitable hymns was likewise provided. The "petition about the widely deplored price increases," the mandate that Stoy refused to read, was a state decree disguised as a prayer, which was distributed to the churches for each minister to read aloud before the actual sermon: "We with our most grievous sins have most surely deserved Thy castigation. We admit and confess it before Thee with aching remorse. Yet we find refuge in Thy great mercy and seek, in the name of Jesus, Thine almighty aid, that by Thy grace the present plight finally may be averted ere we all perish."[20] Public refusal to read a mandate issued by the supreme authorities during a time of crisis was a more serious infraction than mere abstention, and the consequences were foreseeable: Stoy was dismissed from his post as junior preacher in the chapel in St. Walburg.

Stripped of his license to give sermons, Stoy spent three years outside of the church that sources cannot account for. One clue may be found in Stoy's list of publications. In 1773 his "Collection of Seven Hundred Words Taken from the Holy Scriptures as Food for the Heart, the Mind, and the Wit" *(Sammlung von 700 Wörtern aus der heiligen Schrift zur Nahrung des Herzens, Verstandes und Witzes)* appeared; in 1774 his "Observations of a Quiet Soul concerning Help from on High" *(Beobachtungen einer stillen Seele über die Hülfe von oben)* was published. These titles and the professional problems Stoy encountered in

18. Entry in Verlässe der Herren Älteren, rep. 60d, Reichstadt Nürnberg, no. 70, Staatsarchiv Nürnberg.

19. In his "Description of a Journey through Germany and Switzerland, in the Year 1781," Christoph Friedrich Nicolai listed the church registers of Nuremberg (1783, vol. 1, Beylage, 106): There were 1,106 burials in 1770, and 1,833 in 1771. In 1772 the records indicated 1,889 burials. After 1773 the number declined again and began to approach the normal level of around 1,000 burials a year; see also Abel 1981.

20. Mandate of 17 September 1771, A6 Mandat, Stadtarchiv Nürnberg.

the church prompt the suspicion that he was a Pietist and for this reason constantly obliged to defend his character in public.[21] On 25 June 1774 he was finally allowed to return to divine service in Henfenfeld. By this time he was already working on the Picture Academy (see Waldau 1779, 116 f.).

The next eight years were spent in this little village of just seventy-six inhabitants on the feudal estate of the family of Haller zu Hallerstein. There Stoy exercised his religious duties and wrote a few children's books. So this Nuremberger parson's son joined the ranks of theologians in the 1770s and 1780s who left or limited their commitment to holy orders in favor of a career in teaching or writing. Christian Gotthilf Salzmann, Peter Villaume, and Joachim Heinrich Campe were the most famous among them. The careers of these men were founded upon an accelerated restructuring of the church and an unspectacular but steady secularization process in Nuremberg. On 19 April 1783 the last morning mass was held, other religious services deemed superfluous (such as matins and evensong) were discontinued, and in March 1790 private confession and Latin hymns were eliminated during holy communion (see Stadtarchiv Nürnberg 1966, 66 f.).[22] The institutions of the church were not to be abandoned completely—in the first instance, they served to develop a sense of community, industry, and duty. Rather, the faith was to be "tightened up" by the abolition of useless ceremonies and the introduction of modern hymnals.[23] This tighter faith did not necessarily oppose progress or enlightened thought. However, many men trained in public speaking and preaching viewed the extent of secularization of their profession as inadequate and sought another field of occupation in which they could hope to have a greater effect on human virtue: the education of children.

Upon embarking on his teaching career in 1782, Johann Siegmund Stoy not only broke away from his own past but also from a long family tradition of ecclesiastical calling. His father, Magister Johann Friedrich Stoy, and his

---

21. Whether there is any connection here to Stoy's membership in the Pegnesischer Blumenorden, a literary association to which his father also belonged, cannot be ascertained. In the association's archive in the Germanisches Nationalmuseum in Nuremberg there is no information about Stoy besides confirmation of his membership. This society was founded in the baroque period to foster German language and literature; but it also had a strong bent "toward strengthening virtuous nationalistic and patriotic attitudes" (against Gallomania) and toward promoting moral and ethical behavior, to the glory of God (van Dülmen 1986, 20–23). This pastoral literary society can be compared with the Fruchtbringende Gesellschaft in Köthen-Weimar.

22. In 1781 Nicolai (1783, vol. 1, Beylage, 100 f.) nevertheless still criticized the overabundance of church services in Nuremberg, pompous vestments, and strictly orthodox hymnals.

23. A new hymnal was introduced in Nuremberg in 1791 against hefty resistance from the rural population (see Stadtarchiv Nürnberg 1966, 67).

grandfather Johann Paul Stoy were both stewards in the principal churches of Nuremberg, St. Lorenz and St. Sebald.[24]

Johann Friedrich was a man of learning, who besides his priestly duties also worked as a translator, having learned English as a student during his travels to Holland and England. A father of ten children, only some of whom survived childhood, he probably placed all his hopes on his only male heir, who was meant to follow a course similar to that of his father and grandfather before him. Stoy recalled his father as

> a righteous man and an exemplary cleric—who is renowned for his unaffected, profound piety, who from my earliest youth encouraged me to ceaseless industry, kept me under the strictest discipline, and early instilled in me a taste for reading, useful occupation, and a modest way of life—a good father, who in good time accustomed me to the fear of God, to a great respect for righteous, learned, and noble persons, but also to a revulsion against good-for-nothings and loafers; and diligently impressed upon me [to be] honest and kind toward every man. (Stoy 1789, 3)

Stoy's childhood agrees with the conceptions generally held during this period about education, with diligence and the fear of God deemed the most important virtues one could inculcate in a child; and the strict paternal discipline was surely not limited to verbal correction. Johann took his first years of instruction from his father before he left for the preparatory gymnasium St. Egidien in Nuremberg.

In 1762 Stoy began his studies at the University of Altdorf. This university of the imperial city of Nuremberg was a small, rural, Protestant institution, the only one in the predominantly Catholic southeastern regions of Germany, with only a few students. Its origins lay in a gymnasium founded by the council of Nuremberg in 1526 with the support of the German humanist Philipp Melanchthon.[25] Since city life had a bad influence on the work ethic of pupils, the institution moved to Altdorf in 1575 when a school was built there. Three years later Emperor Rudolf II granted the school the privileges of an academy, and in 1622 it finally became a university.

---

24. "The most senior in service of the eight deacons of the two main or parish churches is called a steward [*Schaffer*], *Dispositor.* All marriage unions in the whole city are administered simply and solely by the two stewards. They generally take over the management of all the parochial affairs, e.g., funerals, etc., and therefore are called stewards, i.e., directors" (Nicolai 1783, vol. 1, Beylage, 100).

25. This theologian and friend of Martin Luther had proposed in his early works a program for elementary education, which was implemented in Saxony when the first Protestant schooling system was established for the general public. Melanchthon was involved in the founding of several universities and schools.

It was the duty of all students to spend some time studying at their native university, so as to ensure continuous attendance.[26] So Stoy matriculated at Altdorf for his first semesters. In accordance with his rank and family tradition, he studied theology under the moderate orthodox professors J. B. Riederer and J. A. Dietelmair, who were known for their conservative theological doctrines and their rejection of church reforms.

"At Altdorf the lectures were held, as was then common, either as *collegia publica, privata* or *privatissima;* they were read in Latin, later also in German, and were sixty minutes long" (Leder 1965, 30). The *collegia publica* were main lecture courses; all students were required to attend them. Advance registration was required for the *collegia privata,* that is, a student had to pay in full or in part in order to participate; as the name implies, they usually took place in the professor's living quarters. The *collegia privatissima,* however, were "restricted to wealthy students or to the professor's favorites" (p. 32). The *Nürnbergische Gelehrten-Lexikon* lists Stoy as attending such private sessions on rhetoric held by the professor of poetry, philosophy, and academic history Georg Andreas Will (Will and Nopitsch 1808, 299).[27]

Rhetoric, or the study of the art of oratory and debate, was considered one of the most important skills to acquire at university. The principal task of a professor, even in the eighteenth century, was the lecture. For aspiring theologians the rhetoric and speech-making exercises were of particular importance, with catechism and homiletics completing these exercises. An Altdorf theology student had the opportunity to give a sermon, whose text had been examined by the professor, in an external church or to teach catechism to children or adults at one of the nearby parishes. Altdorf was the first German university to establish its own homiletic and catechismal seminar. This emphasis on skills of oratory and catechesis was a vestige of the long tradition of scholasticism, and before graduation proficiency in the rhetorical arts was rigorously tested.

In 1765 Stoy publically defended his dissertation ("Virtutum homileticarum facile princeps synkatabasis kai synanabasis") before Professor Georg Andreas Will. The defense of a thesis was made before the rector, the dean of the faculty, and all the professors of the university. It took three hours and was held in one of the Altdorf lecture rooms. "From the lower lectern, the respondent defended a dissertation authored by a chairman who had already graduated,

---

26. See also Bock 1950, 408; the pertinent decree for 1768 is cited in Leder 1965, 60.

27. In contrast to the faculty of jurisprudence at Altdorf, which mainly accepted members of the nobility, students of theology were recruited primarily from the sons of the clergy as well as from the sons of poorer families. They frequently stayed in the college boardinghouse *(Alumneum),* which could accommodate a dozen students (see Leder 1965, 68).

and sometimes developed by the respondent himself, against three selected opponents" (Leder 1965, 35). As a rule the dissertation was written by the chairman, in this case Will, and it was one of his principal means of publicity. Stoy's task was to explain and defend Will's thesis before the auditorium and his three opponents, an organized opposition who sought to confute his arguments. The student was not supposed to present any original theses or content, but rather to show the examiners that he was capable of summarizing a text and its purpose and representing these in argument. His abilities in rhetoric, argument, and memory were at issue here, and it was these that were of value for the clergyman's subsequent homiletic and catechetical duties.

After successfully completing the public ceremony, Stoy traveled to Leipzig in Easter 1765 and matriculated at the university. He spent a year in this city, famed for its fairs, booksellers, and well-developed newspaper circulation. He found lodgings with the deacon of the Nikolai Church, near the cemetery, which was in a part of the city that students could afford (see Bruchmüller 1909, 90). It is probable that he traveled under the protection of Will, who had also studied in Leipzig. Will may have recommended that Stoy attend the classes of Gottsched. Will's own *Nürnbergische Gelehrten-Lexikon* recorded that Stoy's course of study included moral philosophy under Gellert, theological sciences under Hebenstreit, canon law under Hommel, and ecclesiastical history under Ernesti (Will and Nopitsch 1808, 299).

Through Professor Hebenstreit, Stoy gained admission to the university church and could test his performance as a preacher "on Sundays, at the pulpits of the main churches of Leipzig." During his stay he visited Halle and other "main cities in Saxony" (p. 300). He returned to Nuremberg to continue his seminarian studies in 1766. A year later he was accepted into the circle of candidates for the clergy and took up the post of preacher in St. Walburg. Neither his course of study nor his time as preacher and parson adequately answer the question of why he gave up his ministry. The church records of the Henfenfeld parish mention an inheritance that he received in 1782.[28] Was that the only reason for Stoy's change of profession?

The Hypochondriac

In 1789, seven years after Stoy had left his ministry, he found it necessary to make public his reasons for changing his profession.

The rumor was circulating that Stoy's publishing establishment (a shop selling books and children's games) was not successful, that he was in debt, and

28. See Pfarrarchiv Henfenfeld VIII/72, 34; see also the version in Simon 1965, which mentions a more recent dispute with the church authorities.

that he would eventually have to declare insolvency. A few weeks later, Stoy defended himself in a newspaper article. This first refutation was not enough, so after a few days, to save his reputation, he ventured out again into the public eye with the bold headline "The eighteenth of March. A word at this time to all the brethren of Feigel, the murderer and grave-digger, who one year ago this day was lashed to the wheel" (Stoy 1789). He took the anniversary of the execution of the notorious Feigel as an opportunity to address that murderer's "brethren"—a jab at his own slanderers. This last public execution in Nuremberg was the result of a murder case that had caused a great stir (Stadtarchiv Nürnberg 1966, 64, 66). Johann Philipp Feigel's gruesome murder of his colleague Carl Gottlieb Langfritz was so shocking that leaflets depicting the mutilated body along with sketches of the murder implements could be seen on every street. Numerous newspaper articles were written, and grisly street ballads sung, about the murderer's life. The sentence passed on him read:

> Sentence on the misdeed of Johann Philipp Feigel, grave-digger's helper, who for the gruesomely committed manslaughter of his comrade Carl Gottlieb Langfriz, grave-digger's servant, had his limbs broken on the wheel on 18 March 1788, by mercy from on high was taken by death out of his mortal misery, whereupon the body was lashed to the wheel. (*Malefiz-Urthel* 1788, 1)

This judgment was followed by a detailed description of the crime and identified the murder motive: Langfritz was murdered for envy of his material possessions.

Choosing this anniversary for his article was certain to guarantee Stoy a large readership, for this incident was still very much on the minds of the citizens of Nuremberg. He listed many personal grievances of his own about unjust treatment, contending that he too was being envied his social standing, possessions, and success. Those who were slandering him, he wrote, were making themselves guilty of a crime similar to that of Feigel against Langfritz. Stoy's good name was at stake. One slander against him was a "skewed assessment" of his state of health. Here Stoy explained for the first time why the prefecture had accepted his resignation.

In this eight-page diatribe, after first expressing his dismay about the occasion that had prompted its appearance and attacking those who had "sullied" his name, he presented a list of ailments from which he claimed to have been suffering since childhood and which were the source of some of the defamatory remarks: among them were poor vision and weak feet. With these two deficiencies he justified his apparent lack of courtesy (he failed to greet people, because he could not see them) and his constantly "well-heeled" appearance (good footwear, which was generally considered an extravagance, was for him a physical necessity). After this introduction, he separately listed the "main classes of lies" that were being told about him.

Concerning the resignation of his curacy, the "second stumbling block," Stoy averred

> that in the year 1781, *I lost all sleep,* from many various starts and upsets, which were mostly caused by a nearly endless string of unprecedented thunderstorms, conflagrations, and sudden deaths among my friends and acquaintances—for half a year, I had not an hour's refreshment of that sort, the lack of which weakened me so much, that at the least exertion I was beset with constant fits of fainting—That for all the efforts of experienced doctors here and abroad, it refused to get better (indeed, to this very day, few are the nights in which I could boast of two or three hours sleep)—that therefore, after years of mature reflection and a complete want of the powers required for the exercise of my office, I at last found it necessary to take a step that was extremely wretched for me by resigning my ministry. (Stoy 1789, 6, original emphasis)[29]

Insomnia and fainting spells, triggered by deaths, severe storms, and conflagrations, were listed by Stoy as causes of his infirmity. Such complaints are characteristic of the eighteenth century. In a speech before the Royal Academy of Sciences in Berlin in 1810, the physician Christoph Wilhelm Hufeland retrospectively characterized that century as a "nervous period" of "constantly rising perfection in human organization." He identified as the dominant changes since the seventeenth century the "cessation of hexing and demonic afflictions, but in their place (perhaps only under a different name) [was the] universality of nervous disease, hypochondria, hysteria, [and] convulsions" (1812, 17). In 1784 Johann Kaempf explained these nervous diseases as the obstruction of bodily fluids in the lower abdomen. The causes of this obstruction or hardening he ascribed to extreme physical exertion, drinks that were too cold, or excessive heat, but also to "rage, fright, and great terror. For how often has fear not produced instantaneous hardening in the breast, for instance?" (p. 85). Constant sitting encouraged the thickening of the juices, hampering their free circulation and leading to weakness, insomnia, tension and pressure, a counternatural composition of the urine, fainting—to mention just a few symptoms from Kaempf's comprehensive list (pp. 86 f.). Such symptoms and their causes experienced a peak in the eighteenth century. They

29. The first sentence of a review of the Picture Academy from 1782 reads "The Rev. Stoy mentions in the notice on the shipping envelope [of one of the installments] a fatal illness as the cause of the delay of his work" (*Allgemeines Verzeichnis neuer Bücher* 1782, 7:380).

were discussed in autobiographies, case histories, manuals, and "advice to the public," and, taken together, could be described as hypochondria.

Hypochondria itself, however, is not easily defined. In this context the historian of medicine Fischer-Homberger has referred to a "conceptual waywardness" (*begriffliche Verwahrlosung,* 1970, 52). The following statement may comprehend all these symptoms: Hypochondria denotes a feeling of physical or mental unwellness lacking any clear pathological basis.

One of the most frequent diagnoses of the eighteenth century, hypochondria was an illness of fashion, or the "price for the new comforts that mankind had acquired, a punishment for abandoning the simple life (Rousseau), the dark side of civilization, as it were" (p. 39). Such dark sides could be brought on by overindulgence in luxury items such as tea, coffee, and tobacco; also by excessive study, reading, and writing. We are thus dealing primarily with a disease of scholars, which "in the eighteenth century counted as a sign of elevated intellectual status" (p. 41) and served a scholar as a convenient "scapegoat for inadequate achievements . . . and as evidence of one's intelligence, erudition, or even genius" (p. 42).[30]

Stoy's self-account resembled the typical image of the eighteenth-century hypochondriac. The symptoms of his physical weakness and their causes agree with Kaempf's descriptions.[31] His catalog of signs likewise included restless sleep and weakness; insomnia might have been triggered by the shock and terror Stoy experienced from the fire that he said happened right next to his parsonage in 1781; according to the church records of the Henfenfeld rectory, the church belfry burnt completely down in that year. Storms and the loss of loved ones might well have also played an important role in Stoy's life. But his justification for returning to the lay world refers to an idiom of the eighteenth century that appears to have been widely accepted: "What counts among the things people say is not so much what they were thinking, on this side of the argument or that, but what systemizes them from the outset" (Foucault 1990, xv). Here we gain one possible perspective on the surviving documents concerning Stoy's life. The conversion of an event such as the belfry fire into a discourse that corresponds to a generally accepted image of sickness is what Foucault describes as a "systematization," and it blurs the hitherto accepted authenticity of Stoy's utterances.

---

30. In their chapter on hypochondria and Kant the hypochondriac, Böhme and Böhme (1983, 387–426) interpret this "self-inflicted illness" (p. 389) as the struggle between the imagination and reason.

31. Here it is not possible to make a precise distinction between "symptoms" and "causes"; the symptoms that Stoy lists already include his explanation.

In Stoy's apology of 1789, which was followed by a second in 1801, the images of fear and terror are clearly presented. He spoke of many ill-intentioned people, whom he would "smash" for "swarming around me like bees" (1789, 2). "This is the pestilence that is lurking in the dark and that so many people, be they only so-called pranksters, are carrying about themselves; which plague has already taken the honor, bread, and life of many an honest man" (1801, 3). He wrote "trembling and in tears" (1789, 2), and the last section of the text from 1789 poignantly portrayed his fear:

> I would now gladly die, since I have justified myself with these words before my fatherland and my many foreign friends. For—the anxious thought has often troubled me and threatened to worry me even on my deathbed, one day—even after my death a pack of asses and curs, voracious wolves and foxes, may dance upon my grave and sully my ashes with their slaver! Nuremberg the 18th of March 1789. Prof. Stoy. (p. 8)

With such a bone-chillingly dark image of blight and infestation, culminating in a *danse macabre* of brutes, Stoy closed his vindication.

Stoy was not unique in incorporating horrific images and metaphors into his discourse. His purpose—the defense of his name and his honor—points to a deeper layer already identified by Hufeland, for whom the eighteenth century may have brought an end to witchcraft but as a result saw a rise in nervous disorders. Hufeland's comment suggests that hypochondria merely functioned as a linguistic substitute for the former demonic illnesses "perhaps only under a different name." It is now clear that these coded expressions do not simply reflect Stoy's own terror but demonstrate the linguistic transformation of images of fear, which have been cut off from their origins in hexing and "demonic illnesses." With its "dehostilization" of the environment (Bilz 1974, 21), the Age of Enlightenment might have liberated human beings from a fearful and demonic existence,[32] yet older hostile terms survived, which Stoy confidently used for rhetorical emphasis in his appeal: "Now speak, dearest fellow citizens! Have not I deserved, for my fate, which has been sorry in many respects, sympathy, support, advice, and comfort in place of vile gossip and malicious remarks?" (1789, 6). Yet Stoy did not explain the abandonment of his curacy by claiming that the village where he preached attracted storms, but rather that "for all the efforts of experienced doctors here and abroad," his ailment "refused to get better."

---

32. In the same way, a bolt of lightning no longer represented the personification of a god but had evolved into a physical quantity. To measure this and direct it into channels controllable by man was the concern at the end of the eighteenth century: in 1788 the first lightning rod was erected in Nuremberg (see Stadtarchiv Nürnberg 1966, 66).

The prefecture accepted Stoy's official reason for changing professions as adequate. This soon-to-be-distinguished professor of pedagogy, who had come into conflict with the church and was described as a "man of the opposition,"[33] wrote in a notice of 1781: "And if I were able and required to abandon my preaching and other duties, no office would be more preferable to me" than that of a schoolmaster.[34]

In a journal of his travels in Germany, Christoph Friedrich Nicolai described the education system in the city of Nuremberg in 1781. Besides a few charity schools and gymnasiums, there were "former preachers," calling themselves "professors," who "also occasionally hold a few lectures for young people who have left gymnasium and Latin school" (Nicolai 1783, vol. 1, Beylage, 86).

Stoy also planned to found such a school, albeit a more up-to-date establishment than the ones founded by former parsons, stewards, and deacons that Nicolai described. The purpose of the lectures was not merely to enable the pupils to memorize by rote, but to teach them practical knowledge that would be of use in their later lives.

> Indeed, the plans for [a school] have been finished for over a year now and I shall soon send them to press. Up to now I had not yet felt strong enough for this enterprise—also I was always a little anxious, in the face of so many objections, about making known my thoughts on establishing a very small institute—nevertheless I hope that I shall be applauded and gather together a small number of ten to twelve boys of good pedigree for instruction according to the best practical rules of pedagogy. (Stoy 1789, 7)

Nothing ever came of this foundation, and Stoy's ambitions as a schoolmaster were never realized (see Schultheiss 1853–57, issue 4, 6 f.). A year after Stoy publicized this plan, Christoph Büchner founded a private institution of similar conception. It is a matter of conjecture whether such competition was the reason for Stoy's hesitancy, or whether he still did not feel "strong enough" for his preferred occupation. The same may apply to his second choice of profession. Stoy eventually worked in a third area: bookselling and everything attached to that business. Stoy himself called it a publishing house, a pedagogical cabinet, and an education business. A true pedagogue he was not, nor even an educational author: a more appropriate professional designation would be pedagogical salesman.

---

33. Pfarrarchiv Henfenfeld, VIII/72, 16; this was also mentioned by Medicus (1863, 252).

34. Notice of 20 February 1781, Darmstaedter Collection 2d1778(6), Staatsbibliothek zu Berlin Preußischer Kulturbesitz.

2

FROM THE MANUFACTURE OF BOOKS TO THE PEDAGOGICAL CABINET

By the close of the eighteenth century, children's literature was a flourishing branch of the publishing trade. In the secondary literature Stoy figures exclusively as a writer but could also be characterized as a dealer in educational products. He himself used various terms to denote his occupation. In 1784 he complained about the great "burden of private publishing" (1780–84, foreword, 14); in 1789 it was "authorship and private publishing" *(Schriftstellerey und Selbstverlag)* that he was engaged in, which concept he specified more precisely as a "pedagogical press" (1789, 5). In 1791 he introduced a new term, calling his entire publishing program a *pädagogisches Kabinett* (see Stoy 1791).

The Eighteenth-Century Children's Book

Although there were already books tailored to a particular age group by the second half of the eighteenth century,[1] the availability and variety of books increased enormously in terms of the numbers printed, their format, and their intended readership. This literature was distinguished by the fact that its lan-

1. Doderer was the first to undertake a systematic survey of the primary sources in German children's literature in his *Lexikon der Kinder- und Jugendliteratur* (1975–82). Previously, this area had been only selectively treated: cf., e.g., Rebhuhn 1925, Rümann 1937, or Kunze 1965. Ewers (1980) offers an anthology of texts from the eighteenth century with an introduction. In particular, Brüggemann and Ewers (1982–91, vol. 1) include such juvenile literature in their bibliographically oriented work. Grenz (1986) wrote numerous texts from the perspective of the Enlightenment and children's books. Schug (1988) provides a chronologically arranged reference work on children's literature; see also the recent comprehensive work by Wild (1990). On the image within the historical context of educational research, and on book illustration, see Böhme and Tenorth 1990, 210 f.; Keck 1986, 1991; Mollenhauer 1983a, 1983b; Müller-Rolli 1989; Schmitt 1997; Talkenberger 1994; and Wünsche 1991.

guage was adapted for an audience of children;[2] it was a subcategory of an increasing "literarization" described by Grenz (1986, 7), and fictional literature such as plays or novels for children increased in number. As it came to be more widely disseminated, it produced changes in the reading habits and requirements of children, parents, and teachers. For those professions that clustered around such enterprises—pedagogues, teachers, authors, publishers, and artists—this change of expectations had methodological consequences, in particular for the production and design of books. The picture layout in a book, the organization of the subject matter—whether fiction or nonfiction—the comprehensiveness of an individual publication or its physical dimensions: all these questions were the subject of intense debate, along with the child's ability to be educated and to assimilate knowledge.

At the forefront of these debates was concern about the acquisition of reason and virtue. In order to insert "proper thinking" and "proper terms" into the categories of reason and virtue, children's literature drew upon two central concepts. One can distinguish, here, between sensible and conceptual knowledge. In the eighteenth century, sensibility numbered among the roots of knowledge and was associated with immediate perception of an object. In educational terms, appealing to sensibility was justified by the claim that a child at first possessed only sensory or *sensible* understanding, so that the content of knowledge needed to be transmitted via the senses.[3] For most contemporary pedagogues, where the thing to be explained was itself intangible, it could be adequately represented by a picture—in other words, a depiction, rather than a symbol or an allegory. Texts and pictures produced for children reveal a process of simplification of the information to be transmitted, which then acquired its own momentum, transgressing social boundaries in concentrated form.

In this respect, the period in which Johann Siegmund Stoy published his Picture Academy was propitious, because it could benefit from the example of two important forerunners: the works by Johann Amos Comenius and Johann Bernhard Basedow, which it repeatedly cited.

Johann Amos Comenius conceived his *Orbis sensualium pictus quadrilinguis* in the seventeenth century with the intention of linking elementary knowledge with the process of learning to read, within one book. The first edition was published in Nuremberg in 1653. Focusing first on the immediately

2. This period was notable, not for inventing literature specifically aimed at younger readers, but for developing certain characteristics in that literature that still hold today. See Brüggemann and Ewers 1982–91, 2:12, and the introduction there (pp. 1–64); also Ewers 1980, 5–59.

3. A comprehensive historical introduction is provided by Merkle (1983); in this context, see primarily pp. 65–86. For a more detailed discussion of Campe's pedagogical circle Allgemeine Revision, see Kersting 1992, 243–74.

familiar world, the work then gradually expands the young reader's horizon to the distant and unknown. The world is represented as a closed system for children, in words and—what was then a novelty—in pictures. In accordance with Comenius's pansophic system, everything was organized by and understood as originating from God; the book opens with a chapter about God and closes with the Last Judgment. The new element is the transmission of knowledge by sensible means: Comenius assigned substantial importance to pictures (here woodcuts) in the instruction of children. He assumed that nothing reached the mind that had not first been communicated by the senses.[4] Stoy too started with the Creation and, like Comenius, ended the Academy with the Day of Judgment. Here too, the world was presented as a closed system. Stoy did not use an omnipresent God as the backbone of his work, however, but rather the Scriptures as history and not as revelational text. Although all things were explained from a theological standpoint, Stoy explicitly justified this with the argument that catechism was a proper part of a child's general knowledge and that it was therefore an excellent foundation upon which to build. Stoy did not so much offer an order actually found in the world, as Comenius's baroque model suggested, but rather one built upon psychological arguments, which presented knowledge for children in such a way as to facilitate and improve the storage and retrieval of facts. Whereas Comenius provided a single, modest woodcut per double page (a simplicity largely imposed by the limits of the woodcut technique), Stoy, by contrast, compiled an entire book of pictures, relegating the text to accompanying volumes.

Stoy borrowed this design from Johann Bernhard Basedow's *Elementarwerk,* whose title continues: "An Ordered Store of All Necessary Knowledge. For the Instruction of the Young, from the Beginning until the Academic Age. For the Information of Parents, School Teachers, and Tutors. For Use by Any Reader to Complete His Knowledge." First appearing in 1774, this work helped establish Basedow's fame as the founder of an alternative reformist school called the Philanthropin in Dessau. His fundamental work of primary education constituted the essence of Basedow's didactic theory and policy.[5] It inculcates civic virtues as the building blocks of the commonweal. The elements of human knowledge are its foundation. The substance of this

4. On Comenius and his pedagogical and philosophical background, see Schaller 1962; a more condensed description is given by Brüggemann and Ewers (1982–91, 2:433–53). Pilz (1967) describes the *Orbis pictus* in detail, along with its successors and emulators.

5. In contrast to "philanthropy" in the general sense, Philanthropinism was a separate pedagogical movement within the German Enlightenment, taking the name of Basedow's model institute in Dessau. The underlying idea was—not atypically for the Enlightenment—to concede to every child the right to develop and train all his faculties, doing so in the spirit of humanitarianism and

general education is tailored to a pupil's social position: contrary to what is frequently asserted, it was not Basedow's purpose to establish a comprehensive school. From the outset his vision had been that of a small school for those of "good breeding" and a larger one "for the masses" (see Krause 1988, 282). He covered topics that Stoy later also took up, but criticized the use of fables in schoolteaching and supplemented subjects like natural history or mythology with lessons on "the study of the soul" *(Seelenkenntnis),* logic, and grammar. Unlike Stoy, Basedow did not try to construct a completely thought-out, comprehensive system to be memorized, but guided the child reader through the general and sensible basics in their "natural order." The illustrations for his *Elementarwerk* complement these "natural orders" and, taken separately, fill a richly illustrated volume of copperplates similar to the one compiled by Stoy.

The Nuremberg theologian, however, took biblical content as the focus of his work, and here the engravings were not illustrative material but formed the central substance in their own right, meaningful even in the absence of the accompanying text. It was a commonly held view at this time that depictions served a more than merely illustrative function. Above and beyond this, they were held to play a role in the creation of knowledge. For example, in the foreword to his "World History for Children" *(Allgemeine Weltgeschichte für Kinder),* Johann Matthias Schröckh described the function of pictures as follows: they were not just supposed to illustrate the text and refer to it; instead the artist "should start out where the [storyteller] has left off; collaborate with him, or rather, surpass him, by portraying the spirit, life, and passions much more sensibly and movingly than the pen and word can do; and in this way he should make the weak impression that the latter has created strong and lasting" (1786–97, vol. 1, unpaginated).

The Picture Academy has frequently been categorized, with reference to Comenius and Basedow, as a "successor to the *Orbis pictus*" (Dyhrenfurth-Graebsch 1951, 37) and as a "parallel work" (Strobach 1978, 35) to the *Elementarwerk.*[6] Stoy himself made the same comparison in an attempt to repudiate any perceived similarity with Basedow's publication: "The first draft of this work had already been written before the appearance of Basedow's" (1780–84, foreword, 10). Nor did he fail to mention his other predecessor: "In short, the owner of the Picture Academy possesses not just one but nine picture books, from the nine fields on each plate—the equivalent of the beloved *Orbis pictus,*

philanthropy. On this movement and its founder see, in addition to the standard work by Pinloche (1896), also Ahrbeck 1976; Herrmann 1993, 99–119; and Niedermeier 1996.

6. More direct successors would be works that refer specifically to Comenius, such as C. F. Müller's *Neuer Orbis Pictus für Kinder in fünf Sprachen.*

which was so useful in its way, and then incomparably more" (directions, 2). This reference to his two forerunners was aimed at convincing the reader of Stoy's professionalism, placing the Academy within the familiar territory of children's literature. Although the *Elementarwerk* does form the basis of the Academy, and although its very conception was inspired by precursors like the *Orbis pictus,* a simple derivation from these older sources does not place it adequately within its historical context.

Nonfiction Books

Classification of Johann Siegmund Stoy's Picture Academy within the various literary types is complicated by its unique and multifarious construction. It unites different types of children's literature with texts of moral instruction as well as pious and devotional writings; it is simultaneously textbook, spiritual guide, and fable. Nonetheless, the most fruitful categorization for further interpretation is to place it in the class of encyclopedic nonfiction literature for children and young people.[7]

Nonfiction books played a decisive role in the transmission of knowledge. They could be designed as encyclopedias and convey a comprehensive impression of all facts worth knowing, or they could be devoted to a particular subject. But the comprehensive and elementary works decreased in number toward the end of the second half of the eighteenth century. This was provoked not only by a constant increase of information in the sciences in general, which made specialization in different teaching areas necessary, but also the disappearance of the drive toward a comprehensive representation of the world. Where once religious faith and teaching formed an underlying and unifying canvas for knowledge, the children's literature of the Enlightenment was partitioned into individual fields, such as natural history, history, geography, and mythology. In travel accounts, anthologies, and schoolbooks children learned of zebras, the deities of antiquity, and earthquakes. Johann Jacob Ebert's "Natural Philosophy for the Young" (*Naturlehre für die Jugend,* 1776–78), August Ludwig von Schlözer's "Children's Primer of World History" (*Vorbereitung zur Weltgeschichte für Kinder,* 1779–1806), and Gerhard Ulrich Anton Vieth's "Physics Companion for Children" (*Physikalischer Kinderfreund,* 1798–1809) are just a few examples of this specialization trend. To generalize, one might interpret the turn toward nonfiction as a shift from a classical educational tra-

7. Brüggemann and Ewers divide young people's literature into the following categories: entertainment; moral instruction; religious writings; reading and writing primers and grammars, rhetoric, and poetry lessons; and factual and entertaining textbooks (1982–91, 1:41 f.). They place Stoy in the category of nonfiction literature, as does Hruby (1986, 170).

dition to a concern with a "realist" education aimed at utility, sensibility, and order.

Many of the works were therefore richly illustrated or conveyed their subject matter in entertaining realistic formats, like travel accounts, discovery reports, or "children's travels at the schoolbench" (Gutsmuths 1785). Often the content was presented in the form of conversations or dialogs between adults and children. The conversational style could be related to the catechismal method of question and answer and in that case would principally help with memorization. The more enlightened, child-oriented variant consisted of a "reasonable" conversation with a child, in which the teacher surreptitiously took the lead and finally drew the desired result out of the child's own questions and contributions. This technique, also termed the Socratic method or "secret engine of education" (*heimliche Maschine der Erziehung,* Wild 1990, 71), appears frequently in the literature and was one of the defining characteristics of nonfictional texts of the second half of the eighteenth century. Countless children's books were designed to be read by the educator before being presented extempore to the child or discussed in a conversation. In the process, it was crucial not to overwhelm the young mind. Gradual progress through successive stages was essential. In the debate about the arrangement or ordering of the material, the principle of advancing from easy to difficult, from familiar to foreign, emerged by the end of the eighteenth century. Johann Christoph Adelung assigned a fixed sequence for the various disciplines of knowledge in his encyclopedia of 1771 (first natural history, then religion, then the sciences), whereas Georg Christian Raff in 1778 drew an example for the progressive augmentation of the demands placed on a child from the plant kingdom, starting within the immediate scope of his experience (garden plants, vegetables) and proceeding from there to unfamiliar territory (exotic plants). Stoy did not break completely with this approach in his series of pictures, but gave it no particular attention.

Among these nonfiction texts, children's encyclopedias and primers occupy a separate field of their own. The Picture Academy is properly designated as belonging in this field insofar as it claimed to provide children with a comprehensive education and to accompany them up to the age of academic study. It was supposed to lay down a basic foundation of general knowledge. In this sense it adheres to the ancient meaning of propaedeutics, from which the word *encyclopedia* stems.[8] The ubiquitous illustrations resemble the Diderot-style encyclopedia; they not only make reading more attractive to children but also

8. See the comprehensive article by Henningsen (1966) on the etymology and semantics of this pedagogical concept—as he terms it.

act as memory aids. The emphasis on the memorizability of knowledge, and thus the function of such books in developing and enhancing a child's powers of memorization, gradually disappeared from formal forewords. In Basedow's *Elementarwerk* the images assumed the task of illustrating the topics mentioned in the text. To present knowledge in sensible form, so that it was easier to grasp, and to present it in a format appropriate to practices other than reading became the overriding goal of pedagogues. The didactic function of the primer was transformed. The concept of "elementary," formerly understood as relating to the foundations of a science, now meant the first sensory experiences of a child—relating to the practical, the everyday, and the professional domains, which may also be described by the term *realia.* Realia were specific pieces of knowledge that formed the basis of the natural sciences, in particular, as opposed to *humaniora* or the humanities, which comprehended the study of the classical languages and literature. This educational approach, which derived from Wolfgang Ratke's and Johann Amos Comenius's efforts in the seventeenth century, was directed toward practical benefit for communal life (cf. Halle 1779–80). The idea was that "education involves the communication of options to a human being, who perceives himself as a will that sets its own purposes and carries them out in the form of . . . 'usable' things" (Ballauf and Schaller 1969–73, 2:221). This explains the growing importance of book illustrations to address sensibility *(Anschaulichkeit).* Following the reception of the work of John Locke and Jean-Jacques Rousseau, it was no longer merely the development of rational understanding that was decisive, but also, particularly in the case of younger children, the formation of "sensual understanding." This shift was paralleled by a move from enlightened rationalism to Philanthropinism in children's literature. It is significant that, as the literature developed, there was a move away from biblical content and a virtual disappearance of the fantastic or wonderful for a time.

Regarded as dictionaries in earlier centuries, reference works gained new momentum from the demands of professional specialization in the eighteenth century. Two other components are of primary importance: first, "the emergence of the religious ideal of civic morality from Pietism in Germany" and, second, the "influence of French encyclopedism" (Hruby 1986, 164). Pietism had placed "Christian prudence" alongside faith "as its civic, worldly component." Thus the professional ethic developed by Pietism had its moral underpinning, and spiritual belief was infused with secular value, in the form of diligence. Through education, the world of work could be transposed into the child's world by means of realia, and even toys, which were then still considered a pastime, were valued according to their utility and as preparation for work.[9]

The French *Encyclopédie* of Diderot and d'Alembert seems to have had a particular influence on the conceptual design of Stoy's Academy. Thanks to this work, from the mid–eighteenth century, the term *encyclopedia* denoted the "accumulation of scholarly material [and] compilation of all accessible information" (Dierse 1977, 52) and ultimately denoted an alphabetical lexicon.[10] This project formed a complicated cross-referencing network linking all the articles together, not unlike the Picture Academy. Furthermore, the French scholars followed a practice of cross-referencing all the *Encyclopédie*'s articles, so that a net of interrelations was formed akin to that in the Picture Academy. In the French enterprise, that net of cross-references represented a world map or a classification of the sciences, in order to confer a unity on the articles even though they were read in isolation. Likewise by the plenitude of his material, Stoy fulfilled his promise to present something "whole." Nevertheless, even though the author of the Academy suggested that it might "be used as an encyclopedia" (1780–84, directions, 13), these similarities could not obscure the fact that the primary aim of the Academy's cross-referencing system was to train the memory.

This was one of the principal differences of Stoy's work—not only with respect to the incomparably more influential enterprise of Diderot and d'Alembert, but also with respect to other children's works being written at that time. Whereas Schröckh's *Allgemeine Weltgeschichte für Kinder* (1786–97) summarized a single field, Stoy attempted to incorporate the entire store of knowledge of a middle-class child within a complicated ordering framework. In this memory system the participant could orient himself precisely because every piece of knowledge possessed its proper place, in other words, a visualized field on the picture tableaux of the Academy. Stoy did not assume that scriptural history permitted a comprehensive penetration of the world, but regarded it only as facilitating the memorization of all the parts of the Academy.

---

9. On the history of toys as tools of play and learning, see especially Retter 1979, also Vogel 1981 and, specifically for the Nuremberg area, Wenzel 1967.

10. In his introduction, Dierse (1977) points out that the term *encyclopedia* is mostly used synonymously with *compilation* or *lexicon;* see also Kossmann 1967–69, 1554–60. But well into the eighteenth century its propaedeutical meaning still comprehended "connection" and "ordering of knowledge" (Dierse 1977, 2 f.). A more precise definition of this concept is attempted by Henningsen (1966); the contributions to a conference covering various encyclopedias, their design, organization, and communication (Eybl et al. 1995) offers some guidance concerning the history of the term in the early modern period. Darnton (1979) discusses the *Encyclopédie,* especially the history of its publication and dissemination.

The Risks of Private Publishing

A discussion of children's literature from the eighteenth century is not complete without a description of contemporary publishing. That way one can better assess the goals and demands placed on authors who designed and sold books for young people at the end of the eighteenth century.

Stoy initially contracted his Picture Academy out to an established publishing house. The early issues up to 1782 were for sale in the Nuremberg Art Bookstore and Publishing House and were published by Weigel & Werlisch, later Weigel & Schneider—a publisher with a long tradition in the sector of children's books and prints. Shortly after the first installment of the Academy was issued, Weigel's partner died: "I should have long since dispatched the new drafts for the second edition, had the chronic illness and eventual death of Mr. Werlisch not delayed the matter," Stoy wrote to Chodowiecki.[11] Delivery of the Academy was a protracted affair. Just a few years previously, Stoy had published a three-volume work, his *Goldener Spiegel,* at the same press—although his name did not appear on the imprint. The first issue (1778), then still appearing with Weigel & Werlisch, had been published under the name of Johann August Werlisch the Younger and dedicated to Johann Georg Haller, a member of a patrician family. Even the second and third issues, of 1779 and 1781, did not explicitly identify Stoy in the work. The Golden Mirror was a collection of various exemplary tales for children from well-known books like the *Kinderfreund* by Rochow (see *Allgemeine Bibliothek* 1780, 8:161). Similar to the Picture Academy, the Golden Mirror was not written by Stoy himself but drew upon various sources to produce an attractive and marketable anthology of stories, each illustrated with a copperplate engraving. This work marks the beginning of Stoy's career in children's literature. The second edition of the Golden Mirror had just appeared when Stoy started editing the first part of the Academy. He was still in Henfenfeld at the time, and his parsonage was transformed into a Picture Academy "Contoir" (Stoy 1782, 4), where, just like at the more usual agencies of publishing house or commissioned store, orders could be placed and subscriptions and prepayments made for individual issues of his Academy. In 1782, after resigning his office and leaving Henfenfeld, Stoy established his own publishing house in Nuremberg. There his Picture Academy subsequently appeared, with text passages printed in the type *(Sixischen Schriften)* of the council and chancery printer C. F. Six in Nuremberg,

11. Undated letter, Darmstaedter Collection 2d1778(6), Staatsbibliothek zu Berlin Preußischer Kulturbesitz.

to whom he had contracted out the work. Stoy had thus founded his own publishing house and described himself as an "author" (Stoy 1789, 2).

Someone who wanted to publish a book at the end of the eighteenth century had three options. An author could hand his manuscript over to a publisher in return for a contractual relinquishment of his copyright. The publisher then owned the profits gained from the printing, duplication, and sale of the book. Or an author could commission out his work to a publishing agent; that is, the publisher reproduced the text without assuming any of the costs, profits, or losses incurred by the enterprise. The third option was to take up publishing in one's own right. In this case, "the author takes on his own account and at his own risk the steps necessary for technical production and marketing" (Köhler 1896, 36). In the eighteenth century honoraria were low and only rarely could an author afford to depend on them for a living. Compounding this risk was the constant wrangling with publishers about the agreed number of sheets to be delivered, which left an author all the more in the hands of his publisher. The result of these intractable economic and organizational problems was a rise in the number of author-publishers.[12]

The idea of authors' taking responsibility for publishing their own work was not new. Nonetheless, a contract with a publisher did relieve them of financial risk (see Berg 1964–66, 1374 f.). What was involved in founding an author-publishing business was not so much the creation of a big business complete with printing press as the assumption of the full financial risk of producing a book. Such enterprises required a good nose for the market and success in selling a product. Stoy, who kept abreast of published reviews, knew the market well. When he embarked on his Academy project with Weigel & Schneider, Christian Felix Weiße, for example, was enjoying much success with his children's periodical *Der Kinderfreund.* Another publishing house owned by Siegfried Lebrecht Crusius in Leipzig, which focused on children's literature and teaching material, was also doing brisk business (see Hurrelmann 1974, 143–53).

Giving authors and publishers access to this market—and getting a new source of revenue for himself—were what motivated Carl Christoph Reiche to found his Scholars' Bookshop *(Buchhandlung der Gelehrten)* in Dessau in 1781 (see Wittmann 1991, 156 f.). With his assistance, an author could use the

12. Besides various periodicals, such as *Archiv für Geschichte des Buchwesens* and the *Wolfenbütteler Notizen zur Buchgeschichte,* see, for a general overview of the history of the book trade and book manufacture, Goldfriedrich 1909, Widmann 1965, and Wittmann 1991. The last reference in particular provides a detailed bibliography. For the eighteenth century see especially the volume of essays edited by Göpfert et al. (1977) and the study by Rosenstrauch (1986). On the interplay between scientific knowledge and printmaking, especially in Britain, see Johns 1998.

bookshop as a center for distribution without having to enter into any fixed dependency. "A scholar had his work printed at his own expense, consigned it to the Scholars' Bookshop for distribution at commission" (Goldfriedrich 1909, 153), and then shared a fixed percentage of his profits with the retailer and book dealer. Stoy also took part in this distribution system. The Easter and Michaelmas fair catalogs for the year 1782—when the first six issues of the Academy were ready—indicated the following as place of publication: "Dessau, at the Scholars' Bookshop" (Messkatalog 1782). The catalog for the Easter book fair of 1784 also lists the eighth issue as originating from this outlet.

One way for an author-publisher to calculate his total outlay and at the same time finance and publicize a work was to offer subscriptions or accept partial payment in advance. The resulting accurate information about the clientele allowed the publisher to calculate his financial risk, and in return the customers had the pleasure of seeing their names in print on one of the opening pages of the book.[13] The gain in social status had a price for the subscriber, however, in terms of financial risk: if the publishing house went under, he was left with a financial loss. So a part of the risk normally assumed by the publisher was directly passed on to the readership.[14] Sadly, there are no lists of subscribers extant for the Picture Academy, by means of which one might have determined a more precise profile of its purchasers.[15]

The main reasons for undertaking such a project were, on the one hand, the greater influence the author could have over the business and organizational management of his book and, on the other hand, the cumbersome nature of the book trade at the end of the eighteenth century. The barter system—the exchange of books in kind—was clearly inconvenient, and it was a long time before cash transactions could become fully established, owing to the lack of a uniform currency. Another advantage for an author in setting up his own publishing business was the option of catering to a chosen corner of the market: Stoy specialized in issues of popular education. Every item carried by his

13. That Stoy too committed himself in this respect by belonging to a circle of subscribers is demonstrated by the surviving subscription list for the newspaper *Chronik für die Jugend* (see Uphaus-Wehmeier 1984, 186).

14. In 1784 the Scholars' Bookshop was dissolved and its former manager Georg Joachim Göschen reopened business as Verlag Göschen (see Berg 1964–66, 1387; Wittmann 1991, 157). Klopstock brought such subscription practices into considerable disrepute when he was compelled to disappoint the 3,480 subscribers to his *Gelehrtenrepublic* (see Ungern-Sternberg 1980, 183).

15. The advance-order and subscription lists for Basedow's *Elementarwerk* were first analyzed by Kersting (1992, 98–113). Despite the similarities between the two works, Basedow's firmly established position in the market, even prior to this publication, allows only limited application of such results to the Picture Academy.

pedagogical publishing house was designed to develop the mental and sensory abilities of a learning child.

From a study of the catalogs for the Easter book fairs of the eighteenth century, it is possible to demonstrate a decline in particular literary forms, such as moral tracts. In contrast, areas such as the natural sciences or education and teaching were expanding, and new book forms began to enter the catalogs. The number and variety of lexicons and teaching texts increased, newspapers and weeklies constantly appeared in the market in updated form, and old educational genres such as the catechism were being squeezed out by the primer. While in 1740 there were seventy-seven religious works as against twenty-three for the fine arts and literature, by 1800 this relation was virtually reversed: twenty-one books of religious content as opposed to seventy-nine publications from the belles lettres (Hiller 1966, 94; Jentzsch 1912). Every year, more than two million books were printed (Raabe 1977, 7), and Stoy saw educational literature rising "like a flood, lapping at the breast of teacher and pupil" (1780–84, foreword, 9). Other theologians, too, such as Joachim Heinrich Campe, Zacharias Becker, Carl Christoph Reiche, and Johann Friedrich Hartknoch, entered the field of education at the end of the century, turning like Stoy to publishing and the production of educational writings. "The clergy were, of course, among the popular authors of the Enlightenment, parsons and preachers, who were accustomed to writing their weekly sermon" (Raabe 1977, 11). They were confronted with a choice as to where more could be done for the enlightenment of the mind: from the pulpit or in the classroom? For many, the decision favored the teaching profession. These pedagogues were interested in disseminating their own educational methods, just as the popularity of the Philanthropinists was supported by their many publications. The educational method that Stoy originally planned to use in his institute in Nuremberg was supposed to incorporate his Picture Academy, and he hoped to exploit its popularity for his proposed private institution.

When Stoy embarked on his publishing career, he had no lack of business: the first issues of the Academy soon sold out. By all appearances, the legacy left to Stoy played some part in helping him maintain a comfortable lifestyle for eighteen years. It was only with the French occupation of the city of Nuremberg around 1800 that his business started to stagnate. His difficulties are typical of the general development of the book trade during the Napoleonic occupation. Repressive measures like censorship, "controls, and levies, even the shooting of the Nuremberg bookseller J. Ph. Palm" (Vogel 1978, 118) were the order of the day—although they were directed at political dissent in the occupied territories rather than at the publishing trade as such.

Even in Nuremberg, traditionally an important center for trade in books and works of art, the number of publications and art bookstores declined toward the end of the eighteenth century after an initial upturn. Even well-known publishing houses, such as those of Trautner or Endter, dwindled to no more than a "better-quality stationer's" (Reynst 1962, 23). "The lesser author-publishers gradually vanished from the scene." Before that time, however, Stoy was able to rack up good sales for his Academy. By the completion of the sixth installment in 1782, the first edition was already sold out, and he had to begin a reprint. By 1789 even the second edition was exhausted. Stoy could no longer keep his promise to acknowledge all his subscribers in print: "Not a third of the owners of the 1,200 copies, which have . . . already been sold, are known to me" (1780–84, foreword, 14). This was because some of the orders came from abroad and many copies were sold through book dealers.[16] By March 1784, twelve hundred copies of the individual installments of the Picture Academy had found a buyer (p. 14).

In contrast to the foreign buyers, the *Mittelsmann* in Nuremberg, for whom the Academy had been conceived, showed no great interest. "It had to wound me sensibly throughout, of course, that my writings should count least of all precisely in my fatherland, that hardly a hundred of my dear fellow citizens know of my publishing house and would rather lay out their money on pathetic daub and drivel than get their children something reasonable and pleasing" (Stoy 1789, 5). In the imperial city of Nuremberg this lack of sales did not lie in any particular aversion to the content and biblical focus of the Picture Academy. It was its reformatory intent of propagating education by means of pictures that was more likely to meet with resistance. The lack of interest among his fellow citizens may be ascribed not least to Stoy's quarrelsome character.

The Picture Academy sold successfully nonetheless. The reason lay not so much in the conditions of the book market as a whole, as in Stoy's 1791 advertisement for his well-equipped pedagogical cabinet. By the time this promotional leaflet appeared, he had greatly expanded his selection. His enterprise could no longer simply be called a publishing house but was a true education business with a "display of goods" (*auslage von waaren,* Grimm and Grimm 1984, 25:711). Books made up the smallest part of his stock. Educators repeatedly complained that there were no suitable teaching materials or good schoolbooks and called for this shortcoming to be addressed. Stoy's education

16. One reviewer for the *Allgemeine Bibliothek* (1782, 10:465) reported a foreign sale to Vienna. On this distribution basis it is conceivable that a large proportion of the copies were sold in rural areas and used by teachers there as visual teaching material.

business supplied such teaching materials. It was among the smaller enterprises that expanded into the domain of toy- and bookselling at the end of the eighteenth century. "If centralized fabrication by labor within the workshop is the trademark of manufacturing" (Kellenbenz 1977, 330), then the publishing enterprise constitutes a domestic precursor for this type of business. In this sense, within his publishing house Stoy combined many other occupations besides publishing, including authorship, bookselling, and retailing. As a manufacturing enterprise, however, his business was also characterized by a growing division of labor. He did not run his own press; he employed draftsmen, engravers, and occasionally women to color the prints, and sold "brand-new articles, never before seen" (Stoy 1791, 3). An enterprise that began as a private publishing house had now entered the educational and toy trade.

The "Education Business" and Its Pedagogical Wares

In 1789 Johann Siegmund Stoy entitled his business "my pedagogical publishing house covering some fifty articles" (1789, 5).Two years later these "articles" included the ones listed in a mail-order catalog bearing the title "Extensive description of the pedagogical cabinet, which I have arranged for the alleviation of education and for the instructive employment and entertainment of the young." In a preface he mentioned a "mail-order enterprise." This illustrated catalog mainly addressed Stoy's "friends abroad" and in the first few pages established how and in what form orders could be placed. Stoy listed exchange and conversion rates, rebate conditions, and shipment costs. These provisions were followed by individual descriptions of the items, each of which was numbered—altogether there were seventy-two little boxes, including alphabet sets, paint boxes, miniature cabinets of natural history specimens, plaster models, building blocks, and a selection of portrait medallions. At the beginning Stoy listed the educational texts, catalog nos. 1 to 21. These were not texts in the usual sense, however, but boxes, folders, and cases: articles indirectly involved in learning the German language. Books formed the smallest part of the selection in any event: there were just four different titles in all, as opposed to four text plates, ten picture plates, seven alphabet boxes, twenty-two other boxes of various designs, twenty-two boxes of building blocks (which he called apparatus), and two games. The items were mostly made of paper or cardboard and wood, with the addition of pigments in the paint boxes. Stoy's establishment could be termed an "education business" (Roth 1800–1802, 3:156), and he himself a "tradesman of manufactured goods." He fits the contemporary description of a merchant involved in the "trade of manufactured goods" or "the storage thereof" who accepted commissions (ibid.).

Stoy never tired of emphasizing that his articles "will certainly satisfy the desires of all, and will offer everything that can serve the education and useful entertainment of the young" (1791, 3). The "useful entertainment" that Stoy wished to offer was one of the most frequently used terms of the time to characterize the process of pedagogical instruction. "Usefulness" could be involved where knowledge was conveyed as well as enjoyment: a miniature model of an ear that could be taken apart aroused children's interest while teaching about the inner construction of the organ of hearing; a miniature shop served not only as an attractive game but also as an instructive object of demonstration. In this way, pedagogical instruction attempted to translate everyday experience into objects that were always ready to hand as a substitute for experience. Here childhood was not merely defined through the experiential and professional world of the adult and in terms of the effort to grow into that world. Instead, childhood became a bounded space in which the appearance and events of the world were mediated. Like all who possessed pedagogical ambitions in this period, Stoy too advocated direct observation on the spot: if a blacksmith were the subject of a lesson, one ought to visit the smith in the city or in the countryside with the children. But this part of an education grounded in real life was not always attainable. Its place could be taken by teaching materials and playthings.[17]

The general growth in the production of books and especially of children's books was based among other things on the demand for materials that were easy to grasp while being both educational and entertaining. Comparable with the articles in Stoy's "education store" is a category of children's books that appeared on the market at the end of the eighteenth century under the title of "occupation books." Among other things they contained instructions for children's games, question-and-answer games, directions for botanizing, and directions for planting one's first garden. The recreational aspect of the time that children were supposed to devote to these activities is readily acknowledged in the subtitles. It must be time sensibly spent, however; "Little Workbook for Children in Their Hours of Leisure or Pleasant and Useful Employment for the Young" (*Kleines Arbeitsbuch* 1810) is one title encapsulating the common

17. An "education shop" like the one run by Stoy was described by Basedow. The accompanying engraving shows the storefront, and one can discern numerous servants and parents carrying out of the shop instruments, models, and natural products, which they have just purchased (Basedow 1771, 200 f.).

program of all such activity books: have fun while learning, never be idle even during leisure time, and choose a way to find out even more.

One of the few works to admit its purpose at first glance is entitled "Occupations for the Youth of All Conditions, to Accustom Them to Purposeful Activity, Amusing Entertainment, and Stimulation of the Taste for the Arts and Trades" (*Beschäftigungen* 1834–40). Purposefulness, utility, and memorization of the material are key in this description. An occupation book used text and picture to introduce the activity, which lay in a median zone between playing and preparation for a profession. The respective authors aimed at a style that was appropriate for children, in this case a gamelike format, and content oriented toward the parents' professions.

In this case, and unlike Stoy's wares, the instruments needed for the games or projects were not included but had to be constructed or bought independently. One of the first books to offer the materials for these useful games was Georg Heinrich Seiferheld's "Collection of Amusing Electrical Devices for Young Electricians" *(Sammlung electrischer Spielwerke für junge Electriker).* It was directed at the interested layman without electrical knowledge and was aimed less at the "individual thinker" than at the curious spectator. The intent of this book was to encourage experiment, since "from children come people, and amusing devices often give occasion for greater things" (1787–99, 2d installment, 8).

In 1784 only the smallest minority of children ever received formal education, and compulsory education was not yet universal at the beginning of the nineteenth century. Even so, Stoy's "education business" and occupation books such as Seiferheld's attracted public interest. By the end of the eighteenth century, toys were being mass-produced for the first time. Like literature for young people, toy production was, from the outset, an educational enterprise as much as it was a manufacturing industry. Even in Nuremberg, known for toy production since the fourteenth century, the output of toys rose rapidly.[18] Since toy making was an "independent art" open to anyone (as far as simple toys were concerned), the strict guild regulations in Nuremberg led many journeymen to seek work in this sector (Wenzel 1967, 82–84). The economic, social, and historical components of rising demand, an opportunity for unemployed journeymen, and the Nuremberger toy-making tradition combined to create a

18. The negative aspect of this economic upsurge is described by Retter: "Contrary to the ambitious tradesmen, toy producers barely made a living under oppressive social and economic conditions. This applied particularly to the cottage industry in rural areas but also to many Nuremberg manufacturers of mass-produced toys, who fell increasingly into the hands of tradesmen who were forever cutting prices" (1979, 76).

new form of trade. Publishers and retailers had always offered toys on the market. With growing competition, however, they were forced to devise new sales strategies. Stoy and another pedagogue were among the first in Nuremberg to do so: Georg Hieronimus Bestelmeier began in 1793 to publicize his program under the heading "Pedagogical Store for the Instructive and Pleasant Entertainment of the Young" *(Pädagogisches Magazin zur lehrreichen und angenehmen Unterhaltung für die Jugend).* Johann Ferdinand Roth reported that "Bestelmeyer's establishment also has in its publishing shop, among other things, a store of useful playthings for the young, which contains many architectonic, mathematical, and physical objects. A printed list in six categories, including prices and copperplate engravings, is available" (1800–1802, 3:127). A history of Nuremberg's commerce (Roth 1800–1802) specifically mentioned its mail-order catalog. This type of distribution, in which the publisher took over the goods from the manufacturer and arranged for their sale, could be used to increase the sale of toys, aided by brief descriptions of the articles and accompanying pictures—an early form of the warehouse catalog. Such catalogs replaced pattern books of textile samples, which were also now illustrated; compared to pattern books, the catalogs had a larger range of goods and additional descriptions.

Comparing the three programs, Seiferheld's stands out as the most specialized, particularly in natural history. Bestelmeier offered, besides building-block sets like Stoy's, physicomathematical, hydraulic, and magnetic objects. Overall, he offered the most curious goods: a model of an ear that could be taken apart, a miniature printing press and theater, a variety of optical magic boxes. Stoy offered the most elementary selection for younger and older children alike. His playthings not only demonstrate an underlying educational model, but also provide insight into the material, technical, and economic conditions of their manufacture. Next to wood and metal, paper was by far the preferred medium of modern times and particularly so in the eighteenth century (see Retter 1979, 56 f.; Vogel 1981, 38–64). It was one of the cheaper play materials, easily cut, pasted, folded, and propped upright.[19] Like Stoy, Friedrich Justin Bertuch also urged that the illustrations of his *Bilderbuch für Kinder* not just be looked at: "The child should be allowed free rein in handling it, as with a toy; he should draw pictures in it at any time, he should color it; yes, with the teacher's permission, even be allowed to cut out the pictures and

---

19. For the uses of paper in play, see especially Blasche's *Der Papparbeiter* (1797), which offered detailed instructions for working with paper and described a useful occupation for leisure hours that yielded many benefits, including mental agility, manual dexterity, and stimulation of the child's imagination and eye.

paste them on carton lids" (1790, 3). The development of educational materials demonstrates that besides the appropriation of the world by reading there was an autonomous appropriation supporting, among other things, the motor skills of the child. The book medium, hitherto a purely passive tool, could itself be an instrument of play in the education of children.

The professionalization of Stoy's business is evidenced not only in the decorative and skillful design of his mail-order catalogs and the use of materials typical of the period, but also in the various advertising strategies he employed to promote the sale of his Picture Academy. The first installment had a foreword by a member of the Erlangen Consistory, Georg Friedrich Seiler, which emphasized the value and practicality of the work.[20] Taking up this point, Stoy noted a deluge of literature in which the reader was barely able to find his way. Thus he was proposing his Picture Academy as a guide, since it contained only the best passages from a choice selection of educational literature (1780–84, foreword, 9). He spoke of having reduced nine illustrated works to a single publication—knowing that most people could not afford to purchase a proper children's library (see Promies 1980, 782). This able marketer deftly intermingled the individual items in his "pedagogical cabinet": for example, the index of his children's Bible also referred to his Academy. Nor is it a coincidence that the children's songbook contained exactly fifty-two songs, for the Picture Academy too was structured around the number of weeks in a year.[21] Besides giving generous discounts, Stoy also knew how to stagger his offers, bringing out variably priced editions of a single item: bound and stored in a slipcase or still in unfinished sheets, colored and offered in a special box, or uncolored and printed on cheaper paper.[22] Item 6 alone, a "Little Biography for Young People" (*Kleine Biographie für die Jugend,* Stoy 1788), could be ordered either as individual sheets mounted on cardboard, with or without cover, or as unbound, loose sheets. Different prices and choices allowed the pedagogical

---

20. From 1775 Seiler was *Konsistorialrat* in Bayreuth and education official *(Dezernent)* for the whole principality. In 1788 he was appointed professor of theology and was simultaneously invested with the offices of superintendent, first preacher at the city church, and headmaster of a gymnasium. Stoy had thus asked an influential man, himself the author of numerous books on biblical subjects for children and young people, to write the preface for the Picture Academy. An idea of how this might have looked can be gleaned from Seiler's preface to Ernesti's "Practical Instruction in the Sciences" (*Praktische Unterweisung in den Wissenschaften,* 1778).

21. Here Stoy conceived his Academy as fitting within the tradition of devotional literature, such as Johann Hübner's *Zweymal zwey und fünfzig auserlesene Biblische Historien* of 1759, whose two volumes were each divided into fifty-two stories that followed the church calendar.

22. This tactic is commonly used on the book market. The Leipzig publisher Reich had his new releases printed on both cheap and expensive paper, in order to reach as many social levels as possible (see Rosenstrauch 1986, 44).

cabinet's range to appear richer and more varied than it actually was; setting aside the many variants, one can count around thirty different books, boxes, and games, rather than the seventy-two listed.

One last example will emphasize the commercial character of Stoy's enterprise. In a first public introduction to his Picture Academy in the form of a printed booklet, Stoy sought to attract another class of customers. Not only children were to be served by his compendium but also artists, such as engravers and painters, who with the plates to the Academy would acquire a set of originals from the greatest German artists, collected from all works of art and "drawn and engraved with all the refinement of Berlin taste" (Stoy 1779, 31). Johann Georg Sulzer's recently published two-volume *Allgemeine Theorie der schönen Künste* had asserted that

> to have at hand a collection of the best copperplate engravings of [works of art by] those artists by means of whom art has truly advanced or been perfected, made by a good master or connoisseur, would produce a most important advantage for the learning of art. This collection should have on every page something new that had been thoroughly accepted in the current state of perfection of art. (1771–74, 2:637)

Stoy's teacher Will also had formulated this suggestion in his 1766 work "The Greatness and Variety in the Kingdoms of Nature and Custom, Linked to the Age-Old Design of the Creator, Presented in a Hundred Fine Copperplate Engravings and Discussed in as Many Pretty and Moral Tales according to My Taste, Published for the Use of Every Man, but of Particular Profit to Young Poets, Painters, and Artists." Even though Sulzer's passage was used here as an advertising ploy and Stoy's contributions could not meet Sulzer's requirements in the least, his endeavor reflects the changing role of pictures and illustrations at the end of the eighteenth century. They became a part of the representation of the world, which could influence human ideas and knowledge and therefore should correspond as closely as possible to reality; that is, they should correspond to the best artistic models of the real world.

What is certain is that Stoy's Picture Academy sold well in the first two installments. The pictorial materials and their associated text formed a single entity for the buyer. Stoy was in a position to have a catalog printed and could stock a varied array of wares on his premises. Upon the appearance of the Picture Academy from his private publishing house in 1782, Stoy had several years of financial security. Whether this was attributable to Stoy's inheritance or to his making a name for himself in the book and toy markets is unclear. What is certain is that his fame as compiler of the Picture Academy extended well beyond Nuremberg and secured his business for some years. With the turn of the nineteenth century and the arrival of the French in his city, however, Stoy's publishing "came to a standstill," and two years later circumstances drove him to offer some of his private paintings anonymously for sale, because, as he justified it, one "is trying to sustain oneself in these wretched times."[23] He was forced to shut down his publishing house and died impoverished in Nuremberg in 1808.

In considering its initial modest success, one wonders why the Academy did not reach the stature of Basedow's *Elementarwerk* or Bertuch's *Bilderbuch.* There are many possible reasons for this. Certainly the smallness of the Academy's pictures is one of them, along with its claim to comprehensiveness, so that although the work sold, it was not used in the manner Stoy had intended. Another reason will be examined in the following analysis of the material content of the work.

Finally, it ought not to be forgotten that the Picture Academy appeared just ten years after Basedow's *Elementarwerk* and ten years before Bertuch's *Bilderbuch* (1792–1813). In emulating the *Elementarwerk* and thus appearing less original, not being the first book of this type on the market, the Academy was too conservative in design and failed to satisfy new requirements in the way that Friedrich Justin Bertuch did with his different conception. In his preliminary announcement, the editor and publisher Bertuch claimed (1790) that a child should be able to amuse himself with the pictures and ought to be in a position to play with them (which Stoy also advocated in his conception). With this aim, the picture plates appeared in a steady stream of individual installments from 1792 onward, accompanied by an explanatory text. However, Bertuch's work abandoned the complicated cross-referencing system of Stoy's pictures; each depiction is individual and described independently from the rest. The colored plates are also detailed, but easier to grasp. Stoy's Academy was not as flexible or comprehensible; Stoy has therefore remained sandwiched between two more famous didactic protagonists of his time.

---

23. Letter of 21 October 1802, Autograph Collection OH, Zentralbibliothek Zürich.

IMAGE

3

SOURCES AND STRUCTURE OF THE PICTURE ACADEMY FOR THE YOUNG

Johann Siegmund Stoy conceived his pictorial atlas in a period when the function of the image was changing. If the image in the pedagogical context had previously assumed an allegorical, symbolic guise within the rich moral and didactic canon, during the eighteenth century this allegorical function was lost, and it came to be associated with the direct exercise of sensibility.[1] Although Stoy's solution attracted enough customers for his schoolbook, his project increasingly dimmed before Bertuch's enterprise at the beginning of the nineteenth century. Stoy's dilemma is best understood within the context of the theological use of the image, the tradition that he inherited.

Within the Tradition of the Biblical Image

In performing his didactic and persuasive duties, the medieval preacher used image and word to disseminate to the people the Christian faith, the divine image, and the notion of the saints. The image was particularly important in supporting the tasks of memorization involved in this process.[2] Church buildings provided the most imposing medium for the religious image: wall frescoes, altarpieces, and stained-glass windows. The "portable image," in the form of pamphlets, broadsheet woodcuts, pilgrimage attendance rolls, and, later, emblems in edification literature (the *Erbauungsbuch*), developed toward the end of the Middle Ages. Only in the course of the seventeenth century did pictures, along with textual material, eventually begin to appear in the homes of those for whom they had been initially designed—the uneducated at the lowest echelons of society. Thus religious depictions shared with the sermon,

1. See Harms 1990, 241–324, and Warncke 1987, 137–216, for more details.

2. Ringshausen (1976) traces the development of the biblical image from book illustration to teaching aid. The attitudes held in early modern times about pictures are described by Warncke (1987) from the point of view of the history of media; this study pursues the connection between word and image during that era. The picture's place in mnemotechnics (see p. 137 below) before its "revamping" *(Umrüstung)* in the fifteenth century is described by Berns (1993).

the spoken word embellished by rhetorical gesture and expression *(Mimik),* the most important mission of spreading the doctrine of salvation.

"The *Biblia pauperum,* first appearing in the thirteenth century," which displayed "the essential facts of salvation in pictures, was primarily a teaching tool for the *clerici pauperes*" (Matuszak 1967, 13)—not, as its name seems to suggest, a pictorial representation aiding the lower orders to gain a better understanding of Scripture, but rather excerpts from the Bible intended as source material for sermons and instruction given by the *pauperes,* men and women living in devout poverty. Characteristic of this pictorial program was that the individual had to possess a large amount of previously acquired knowledge in order to make sense of the pictures. This was particularly important because many of the visual forms were supplied with inscriptions in Latin. Besides these *Biblia,* whose extracts were selected with a view to fighting heresy and promoting moral theology (Weckwerth 1979, 8), a sort of medieval edification book, the *Speculum humanae salvationis,* appeared at the same time. This mirror of redemption, as the *Speculum* was described, summarized various exemplary stories, and whoever "read in it, would simultaneously be looking in a mirror" at their own image (Matuszak 1967, 8). The *Speculum* served a function similar to that of the *Biblia pauperum* but supplemented the depiction and narration of biblical events with encyclopedic knowledge, exemplified in the three *Specula* by Vincent of Beauvais, dating from the thirteenth century. In these works general knowledge about natural and historical matters is woven into the narrative of biblical history or used as an illustration.

Specula and *Biblia* may be considered the most important typological works of the Middle Ages. In both forms the relation between text and image established the preacher in a mediating position, a teacher who explained their significance orally to the novice. Unlike visual depictions within church buildings, these works always required the spoken word and were associated with the instruction and informing of prospective clergymen or preachers from poor backgrounds. Hence they were pictorial and verbal digests of the essential rudiments of Christian doctrine. The image and the speaking interpreter together assured that the compilatory works were properly understood, and during this presentation the faith might be related to the world of human experience and everyday life.

Until the seventeenth century, exemplary and edification literature, sermon parables, and catechisms also used pictures predominantly as visual support for oral transmission of the faith. Embedded within allusions, reports, and comparisons, they drew thematic connections to the everyday experiences of members of the congregation. Edification literature can be understood as a collective term referring to all writing demonstrating a concern with the condition of the soul, containing some form of textual interpretation, and intended to lead to practicing Christianity. Therefore, it always provided specific guidelines on how to act and took an educative stance. Pictures frequently accompany this literature as well. One of the most famous sources of moral improvement from the seventeenth century is Johann Arndt's "Four Books on True Christianity" *(Vier Bücher vom wahren Christentum),* which were successful not least because of the emblematic manner in which they were illustrated (see Peil 1977–78). Essentially, these tracts served as lifelong memorization aids for their intended readers and accompanied a family throughout the entire year. Consequently, this one book was read over and over again like a catechism.

> Edification literature, basically comprising all Christian literature concentrating on piety as the entirety of religious behavior, both within the community and within the personal domain, received an impulsion toward piety stemming from theology; on the other hand, it also encompassed a critical momentum associated with an affective and emotive counterweight to theology. (Mohr 1982, 43)

The visual image in edification literature may, by its sheer directness, trigger skeptical impulses, whereas the text may prompt such a response by the emotionally loaded models it recounts, which are supposed to teach a rather abstract dogmatic lesson but draw their material from wondrous miracles and curious facts. The official church was constantly at risk of seeing its teachings made too simplistic or worldly, and its use of the vernacular made it that much more susceptible to such criticism.

The various Christian picture books or illustrated Bibles of the seventeenth century should feature in any analysis of the religious image predating the eighteenth century.[3] The illustrations in Sigismund Evenius's "Picture School" *(Bilder Schule)* are remarkable not just for their Bible-like allegory but also for their function as memory images and hence their turn toward a form like that of a mnemonic work. Evenius was the first to propose that pictures also be used as visual aids in the classroom, inspired by the suggestion in Campanella's *Civitas solis* and Andreä's *Christianopolis* to exhibit picture plates for teaching purposes. Evenius, Johann Arndt, and Johann Amos Comenius more or less strongly advocated the use of pictures. Underlying their statements was the confidence that through an observational retention of word and

---

3. A comprehensive listing of early picture Bibles, as well as picture catechisms, is provided by Brüggemann and Ewers 1982–91, vol. 2. For the role of the pictorial image in religious instruction over the centuries, see the study by Ringshausen (1976). See also Oertel 1977.

image a pupil would perceive "behind the statements the greater whole and its purpose" (Ringshausen 1976, 78).[4]

In summary, within the theological context of images up to the eighteenth century, three main properties or functional and assimilative methods emerged as important defining elements of illustrated children's literature in that century: The comprehensive illustrated book might be a lifelong companion to its young reader as a reference work; like the Bible, edification reading or the Picture Academy also contained everything a nonscholar should need to know throughout his life. The pictures in such books could attract the viewer's interest and at the same time fix the depicted material in the memory via text and image. Finally, the image needed a mediator, an oral interpreter, who could form a bridge between the depiction and an uninformed person. These three central aspects, a universal book, memorizable images, and a mediator, are all represented in the frontispiece for the Picture Academy's fifty-two elaborate copperplate illustrations.

Clio and Minerva: The Introductory Plates for the Academy

On the Academy's frontispiece we see Clio, the Muse of history, sitting on a cornerstone block in the middle of the scene (plate 1). Her left hand is resting on an open book, the text of which is visible but not legible. With her right hand she is pushing a curtain aside with a trumpet to reveal a wall of pictures. The scenes in some of the pictures are barely discernible, but Clio is holding the curtain open far enough to show part of five more frames. The skimpily dressed putti standing before the wall of pictures are looking up at Clio or the represented scenes. One putto, seated on the block behind the Muse, is trying to sneak a look at the pages of text in her lap. In the foreground, some books (Basedow, the Fables, and three volumes whose spines read "the Arts," "Natural History," and "Education") are strewn at her feet by two other picture plates (Apollo and the Muses, a tower). The wall illustrations that Clio is unveiling contain scenes from the Old Testament (Adam and Eve, Cain and Abel, the construction of the Tower of Babel) set beside other subjects linked by the Academy to these biblical themes (Romulus and Remus, and the construction

4. Johannes Buno, another pedagogue and theologian of the seventeenth century, drew his picture Bible more closely toward the *ars memorativa* than the above-mentioned clergymen did. His readers get a pictorial introduction to each Bible story, which they may memorize, chapter by enumerated chapter and verse. Another source of inspiration for Stoy might have been Christoph Weigel's *Welt in einer Nuß*. See, e.g., Schug 1988, 319 ff.; on Buno's method in particular, see Strobach 1979.

of an Egyptian pyramid). The cornerstone around which the group is gathered bears the inscription "Clio gesta canens transacti temporis edit" (Clio descants upon the deeds of the past).

This stylized classroom scene anticipates the Academy's method of communicating knowledge. By means of pictures, common analogies, and an oral presentation by an instructor, the class may, by attentive observation, learn about biblical Scripture and world history. The child beside Clio fails in his attempt to peek at the book, which would be undesirable. So this Muse of epic narrative and daughter of Mnemosyne recounts stories of the past to the children.

The Muse of history is a favorite allegorical figure in eighteenth-century publications. Her characteristic attributes—a book or scroll, and a stylus or quill—identify her as the memory of mankind, recording and storing what she observes in an effort to preserve human and natural history and with these writings to give people a morally instructive mirror. In most cases—as in this frontispiece—she is seated and is often paired with Minerva, the tutelary goddess of the arts and sciences (Minerva appears in the Picture Academy's dedication plate).

The work's title, Bilder-Akademie, is blazoned across the central draping fold of the curtain Clio is lifting. Stoy's purpose is not a "picture school" of the type Sigismund Evenius proposed in 1636 with his *Bilder Schule,* a children's introduction to "godliness," the religious and preceptoral character of which the many illustrations were supposed to demonstrate, but rather an academy of pictures. Since the sixteenth and seventeenth centuries, the term *academy,* which goes back to Plato, denoted a form of conservation and generation of scientific and artistic knowledge. The union of science and technology took precedence over any religious focus. *Academy* was already appearing in book titles during the sixteenth century but flourished in the seventeenth century and particularly in the eighteenth, to the point that it was used—as in this case—even for children's books.

In the eighteenth century an academy was an institution for teaching, collecting, and assembling knowledge, where inventions were made and research conducted. It could take the practical form of a noble's *Ritter-Akademie,* or the scientific one of the Royal Society.[5] While in French-speaking regions the term *académie* clearly indicated the orientation toward research, in Germany it was a foreign word and not clearly distinguishable from the idea of a university. Only gradually did it become established (see Dickerhof 1982, 42). As

5. The concept and realization of the scientific academy in the eighteenth century is surveyed by Hammermayer (1976) and Hartmann and Vierhaus (1977).

an example, the development of academies of art in the eighteenth century revealed a struggle for a higher level of learning and a reinterpretation of theory and science. At the same time, however, there was a drive to preserve and convey skills like metal casting and the fabrication of silk. Therefore, the academy was also able to serve the common economic good (see Busch 1984, 177 f.; van Dülmen 1973, 667 f.).[6] Anyone employed in an art or craft that applied drawing skills was required to have had academic training. This requirement had been instituted in an effort to promote and vitalize trade and industry as well as to improve the overall position against foreign competition. A book on husbandry urges an apprentice "to represent in a drawing everything he himself knows about what appears below concerning economic and important matters, the better to instruct those craftsmen and artists, when something needs construction" (*Gründlich- und nützlicher Unterricht* 1720, 11). Drawing was not a purposeless art; rather it was accepted as a practical skill in the service of cameralism and commercial science. "Of note for the eighteenth century is the state's effort to break the guild monopolies by means of the academies" (Busch 1984, 179). Academies exist "so as to elevate the sciences and arts": this unambiguous description in Zedler's *Universal-Lexicon* (1732–54, 1:241) reveals how much the academy concept is laden with utilitarian purpose (see Kraus 1977, 143).

The Picture Academy seems to oblige these interests. Stoy himself stressed how important basic knowledge was for the future practice of a profession. In his outline to the Academy he wrote: "That by this book, as in an academy, under the auspices of good taste and morals, the most noble arts and sciences be united, for the young to be shown all that is beautiful and enticing of further research, is clearly elucidated in the printed notice; and in what form and attitude this should occur, in the finished picture plate" (1779, 4). The epic Muse and the scientific and artistic bent of the academy concept suggest the direction this work took. A similar frontispiece that Johann Peter Voit picked for his *Schauplatz* should help narrow our definition (plate 2). Voit's work, which Stoy also admitted into his Academy, is a compendium of the commercial arts and sciences for young nobles. On the copperplate engraving, the paterfamilias is explaining a wall of pictures to a boy who is facing him. In the shadowy foreground another boy is kneeling on the floor at a microscope. The scene takes place in a room with a very large window with a commanding view onto a well-tended, stylized garden. The window possibly alludes to a division of the lessons into two parts: alongside the preceptor's lessons within the closed

---

6. Schrötter (1908) discusses the painter's academy founded in Nuremberg toward the close of the seventeenth century that J. von Sandrart directed from 1674.

chamber, that is, the theoretical portion of the professional subject matter, is the practical and visual training that may take place only in the craftsman's workshop. The specialized knowledge or practical sciences (such as trade, commerce, mining, naval or military science) must be presented as directly as possible but initially within a closed space.

Following this overture to the Picture Academy is the second of the copperplates bound in front of the actual illustrations: the dedication plate. This expression of the author's affinity with or obligation toward another person goes back to antiquity. With the invention of the printing press and the consequent ability to produce many identical copies of a single work, dedications took on a quite different meaning, since now the name of the acknowledged individual (frequently the financial patron of the work) gained considerable publicity. Only around 1800 did the dedication form seen in Stoy's work change to the one still used today, "To . . . as a sign of personal indebtedness and appreciation" (Schmitz 1989, 237).

The Picture Academy was dedicated to the Swedish crown prince Gustav Adolph, although it is not certain exactly which Gustav Adolph is meant to be honored by it. The sources do not document any financial or intellectual mentorship by the Swedish king. Stoy mentioned in the dedication that his efforts were "but sun motes in the good cause of education" against "what the great educator, *His Serene Highness the Swedish Prince*" had achieved "many years hence" (original emphasis). Gustav IV Adolph (1778–1837, period of reign 1792–1809) was two years old at the time of Stoy's dedication. Presumably these achievements of the royal house of Sweden must be attributed to an earlier monarch. The father of the crown prince, Gustav III (1746–92, period of reign 1771–92), attempted during his years of regency to establish greater social equality among the classes, an end to domination by the nobility, and more religious freedom. As a result, the middle class was considerably strengthened during his reign. Even before then, though, during the Thirty Years' War, another Swedish monarch also gained prominence among the German principalities and free cities. Gustavus II Adolphus (1594–1632, period of reign 1611–32) introduced a series of reforms in his country at the beginning of the seventeenth century, changing the central and judicial administrations, and driving economic development. Unsettled by the Habsburg emperor's growing power, he became embroiled in the Thirty Years' War in 1630. His interest lay in the defense of Protestantism and building up Sweden's power. As a consequence he became a central figure for the Protestant enclave of Nuremberg in 1632 by preventing it from being engulfed by a wave of Catholicism.[7]

7. Reicke 1896; cf. the mention of this monarch's birthday in Stoy's calendar (1780–84, 700).

Exactly what Stoy specifically wished to allude to—whether the king's religious protectorship or his enlightened absolutism—can only be surmised.

It is of note, and speaks for the conception of the copperplate volume as an integral book, that Stoy chose separate engravings for the frontispiece and dedication plate. Illustrated dedications first appeared in painted books during the Middle Ages. The author or translator offered the work to a person of importance—from either the lay or the spiritual world. "Where Christ and the saints were the recipients, an intermediary (intercessor) was occasionally engaged (usually a patron saint) who received the book from the dedicator and forwarded it" (Milde 1989, 237). Stoy took up this tradition of illustrating the dedication page, even though in his century it had given way to epigraphs, forewords, or simple portraits of the honored individual.

The Academy's dedication plate contains a portrait of the infant crown prince (plate 3). A young man is kneeling before this depiction. In his left hand he holds open a Picture Academy, so that the fields or partitions of one plate are visible. His right hand rests on the head of a child who is standing beside him. Other children are settling on the floor around a genealogical plate that they are admiring, presumably that of the Swedish royal family. There is a globe nearby, along with geometrical measuring instruments. Scattered about to the left of the young man, who presumably represents Stoy, are pictures of plants and animals, a painter's palette, a butterfly display case, a writer's quill, an engraver's burin, and some books. These plates, the kneeling figure, the group of children, and the globe are all on the same level. Two steps lead up to a stage farther into the room. Helmed Minerva is standing on the first step, with spear and shield in hand, and looking up at the portrait of the crown prince. Apollo is moving forward next to her on the higher step, extending his hand to accept the Academy from the young man. At the center of the stage is a pedestal crowned by the royal infant's oval-framed portrait, which is held in position by two cherubs. In the background, on the same level as the children but behind the stage, two women are being accosted by a couple of unruly children. To the right of the pedestal, the three Muses are standing together behind the Greek figures. Some female statues fill the niches on the far wall of the room. The aesthetically gathered drape is drawn aside farther than in the frontispiece scene and announces the coming plates. Thus the Picture Academy is presented to a prince, who cannot receive it in person, at the altar of the Muses. Unlike the frontispiece, the emphasis here is less on the medium of learning than on its substance. Natural specimen cases, a globe, botanical plates, and measuring instruments lie on the floor as heralds of the fields of natural history and the vocational trades. While the frontispiece indicates biblical and profane history, here it is the flanking fields, 5 and 6, that are given prominence.

These two programmatic sheets are the introductory images of the book. By choosing an opening plate, which, as a rule, was not frequently done in children's literature, Stoy assigned a special position to the illustration volume. What the title page and dedication text provide in the commentary volumes is visualized here in the form of frontispiece and dedication plate. So Stoy was being consistent, demonstrating once more how he favored looking at pictures above reading.

Text and Illustration Sources of the Academy

The Picture Academy gives the impression of extensive and careful collection by its author and editor. In the second paragraph of his outline to this work (1779) Stoy listed some of the literature he had used for the individual subjects.[8] It is striking how many more sources he used for the fields of natural history and the vocational trades than for the other subjects. Both these themes and biblical Scripture were placed in the largest partitions on the plates, and Stoy seems to have researched them most thoroughly and carefully. The central series on the picture plate, fields 5, 1, and 6, form the axis around which Stoy's Academy revolves.

> I have scoured the whole great stock of educational works that have appeared in the last 15 to 20 years, during which period I was preparing the Picture Academy—and did not shrink from the labor of using the best and most comprehensive works for certain divisions, particularly in the kingdom of nature and the vocational trades, and collecting the best material for primary instruction from them. (Stoy 1780–84, foreword, 12)

Although the Academy's title picture referred more to biblical and mythological scenes, the dedication plate, as the more worldly aspect of the Academy, made direct reference to attributes of natural history and the professions. This emphasis, on worldly knowledge on one hand and Scripture on the other, Stoy made into a central theme, which is the main message of the illustration shown in plate 4. The book of nature and customs stands next to the book of religion. This illustration is taken from copperplate 36, whose source is Basedow's *Elementarwerk.* Stoy borrowed both picture and accompanying text (pp. 720–22). A tutor is standing with outstretched arms at a desk loaded with books and

8. The discussion below focuses primarily on the textual sources of the Picture Academy; for a complete retracing of its illustrations see Reuter 1994.

instruments. He is reaching out to the closest boy in front of him with his right hand, while pointing with his left to some words on the wall: "Book of Nature and Customs, Book of Religion."[9] Behind his back on the wall are plates with motifs from natural history. The boys are standing quietly and respectfully before him. The accompanying text explains that this lesson was "for the virtue and blissful happiness of children" (p. 720). Virtue and bliss—as Stoy and Basedow saw it—were imbued in young minds when their lesson proceeded according to children's own taste, in a natural sequence that they were able to follow, and which helped them to become, "in manhood, of useful and honorable service to human society" (p. 722).

In order to meet this demand, a carefully thought-out teaching program was needed, which Basedow proposed as follows: "All our sciences . . . each one in relation to every other; following a single plan; and following textbooks, each one of them referring to every other; and thus instruction is made *much shorter, more consistent* and *practical*" (quoted from Schöler 1970, 60, original emphasis). One might easily think that Basedow was describing the Academy's structure rather than his own curriculum. But, far more than Basedow did in his *Elementarwerk,* Stoy was able to link the various disciplines and assert that the Academy contained all necessary knowledge. Stoy tried to lend a universal character to it, from the formal point of view, and incorporated all the Philanthropinist's requirements within his own design. It should provide all-round education so as to benefit humanity as a whole; it should not merely teach facts, but in the first place train children how to think, while engaging their bodies by having them replicate for themselves things that were discussed in class, in the form of experiments and work projects. The hand, and especially visual sensibility, were among the most important portals to a child's mind, which could best be accessed with copperplate engravings, cabinets of natural history, and even a *Theatrum naturae et artis.* In Stoy's Academy, these methodological requirements are condensed within the smallest possible space.

Since medieval times elementary education mainly comprised reading, writing, and arithmetic. By the seventeenth century, however, learning as a whole had become increasingly oriented toward encyclopedic knowledge (see Dolch 1982, 267; Schöler 1970, 23). The academy model and the science-oriented *Realschule* expanded at the same time, as a consequence of the birth of industrialization, growing trade opportunities, and a rising middle class. Thus we may conclude with the education theorist Joseph Dolch: "For the primary-school curriculum, the subsequent vocational practice was crucial; for that of the *Realschule,* it was the coming vocational training" (p. 326). The rudimen-

9. The book of nature was a common metaphor for the physical world.

tary knowledge gleaned from the Picture Academy was supposed to constitute an underlying general education, both for vocational training and practice.

Knowledge about nature and the arts and their correlated moral messages are at the center of the Academy's pages, just as they are for the Philanthropinist educational method. It was these messages that helped a young man to become a useful and honorable member of society. Virtuous comportment included veracity, self-discipline, diligence, neatness, and flexibility, while rejecting indolence, immodesty, selfishness, mistrust, and extravagance. Each piece of information contained an exemplary moral. The botanical plates on the wall served an educative purpose while illustrating that even an inconspicuous plant that does not immediately arrest a person's attention belongs within the general canon of living beings and may even provide a powerful cure for human illnesses. Here the child is shown how much external beauty can deceive, while insignificant things may hide many uses. The book of nature and the book of religion are linked by a reference taken from real life, which places the hardworking, active person in the foreground yet links him to a higher truth.

The Field of Natural History

> Field no. 5 belongs to natural science; Ebert and Raff, the *Leipziger Wochenblat,* and the *Kinder-Freund* form the basis of this field; and from these beloved works, what must please the eye and tickle the little ones' fancy has been used. (Stoy 1779, 6 f.)

Johann Jacob Ebert's *Naturlehre* in three volumes appeared in as many editions before 1796 and, together with Georg Christian Raff's *Naturgeschichte,* counted among the most famous natural science books for children in the German language. Johann Christoph Adelung's *Leipziger Wochenblatt für Kinder* was a periodical offering a mixed assortment of fairy tales, stories, fables, letters, and moral commentary along with interesting facts about nature. Christian Felix Weiße's *Kinderfreund* explicitly emulated the weekly but took the form of a moralistic periodical: a fictitious mentor leads a conversation with other people as he guides the reader through the stories.

Stoy made liberal use of these texts as source as well as inspiration for his natural history section. Whole passages were taken over, others were slightly altered, and there are some sections written by Stoy himself. Since some of his sources are lengthy compendia, Stoy was forced to cut down the content so drastically that the sources for passages and pictures can rarely be recognized. Nevertheless, accounts of one particular topic show marked similarities across the board in the children's natural history literature. Stoy attempted to condense to fifty-two plates the same material that fills Ebert's work of six hundred pages. From his commentary we gather that, for him, man is the governor of nature; the natural world, on the other hand, must be regarded as the work and evidence of God. In this way Stoy formed a close tie between natural history and the biblical focus. The Picture Academy presents to man, as lord and master, the three kingdoms of nature: the worlds of the animals, plants, and rocks and minerals. A thematic line binding seashell with exotic herb and ape attempts to teach nature's lessons without, however, being able to substitute a history of the natural world.

In his three-volume children's work on natural science, *Naturlehre für die Jugend,* the scholar and writer Johann Jacob Ebert discussed nature in letter form for his young readers from the *Bürgertum* and aristocracy. While his first two works on natural history (1773, 1775) were directed toward classes in lower-ranking schools and to pupils aspiring toward professions in the arts and crafts, this work on natural philosophy was intended as a guide for parents and private tutors and was supposed to give children between ten to twelve years of age a first taste of science. For Ebert, the essential purpose of the study of natural science was to explain natural bodies, prove the omnipotence of God, and combat superstition and needless fearfulness. Like Stoy, Ebert helped himself to figures and information from various other works (cf. Brüggemann and Ewers 1982–91, 1:1012). He followed a strict classification that essentially adhered to the Linnaean system. In addition to treating the three kingdoms of nature, he also discussed the basic concepts of physics along with electricity, astronomy, and cosmology. His *Naturlehre* contains twenty-eight folding plates bound at the end of each volume. A comparison with Stoy's illustrations and texts on natural history yields many overlaps, some of which are outright duplications.

These overlaps expose the different criteria Stoy employed in deciding how best to incorporate such texts and illustrations in his Academy. For instance, Ebert depicted various primates individually—"Joko or forest pygmy," the "gibbon," "the monkey with the Chinaman's cap," "the great baboon," "the Mone," and "the Loaita"—whereas Stoy threw them together in a single illustration, while retaining their characteristic poses (cf. Ebert 1776–78, 1:358–63; Stoy 1780–84, 590–94).[10] Ebert's "One hundred and twenty-second letter,"

10. The tradition of monkey postures and facial expressions in illustration predates Stoy and Ebert. The "forest pygmy" *(Kleiner Waldmensch),* for example, existed in the form "monkey with a stick" since the seventeenth century. It first appeared in 1699 in the book by Edward Tyson, *The Anatomy of a Pygmy. Compared with that of a Monkey, an Ape, and a Man.* Just as natural history classics by the ancient Greeks were routinely paraphrased until the end of the eighteenth century, so also do the pictorial archetypes recur with the same design elements (see Lepenies 1976, 30).

which provided the commentary on the monkey illustrations, bears the title "The True Apes, Baboons, and Long-Tailed Monkeys." This is one of the few letters that, excepting the salutation, Stoy copied almost verbatim. He merely omitted such finer details as the characteristics of an ape's hind feet, details about the diversity of the species, and quotations from travel accounts. Additions that are not in Ebert's work are generally more of an anecdotal nature: Stoy added to Ebert's two stories a third exciting adventure about how Indians catch monkeys (cf. Ebert 1776–78, 1:592). The article about primates is one of the Academy's longest; and both its length and emphasis on anecdotes reflect the tensions and fascination revolving around the eighteenth-century controversy over the similarities and possible relatedness of apes to humans beings, especially concerning what was, purportedly, the most rationally gifted race of them all, the white European.

Stoy proceeded similarly with the other scientific sections.[11] For instance, Ebert treated the class of birds in forty letters (see 2:1–155), whereas Stoy introduced the general topic of birds in the first plate and supplemented that illustration with three pages of text. He used the same bird classification as Ebert and in later plates (copperplates 23 and 28) discussed individual kinds, which he then correlated thematically—which sets him considerably apart from Ebert. Ebert also wrote about vultures that "can carry off not just sheep, but even calves and ten-year-old boys" (p. 15), or about parrots that can learn to say "not just single words, but even short sentences" (p. 49), but Stoy went one step farther and used the descriptions for his new ordering system. Parrot and vulture return in a totally different context, namely, "animals that can be taught tricks" (Stoy 1780–84, 415–17). Ebert's parrot merely featured as a member of the classification scheme, whereas Stoy's acquired an attribute centered on the human being: a source of entertainment. Consequently, in copperplate 23 the parrot is depicted in a living room. Besides the apes, doves, parrots, and magpies were singled out as teachable animals. They appear mostly in domesticated form and in a context that manifests their usefulness to mankind. Where Ebert frequently depicted them on a blank background, in Stoy's work they are firmly bound into a human setting. The zebra, which for Ebert was an exotic creature, was assigned by Stoy a special place among domesticated horses, mules, and donkeys. The associated illustration shows a man and a boy observing the animals, while their various practical uses are discussed in the commentary. Stoy even priced the zebra: by now, "even in Africa 14,000 ducats are paid out" for it (1780–84, 429), which corresponds to the substantial

11. All Stoy's examples in cosmology and astronomy, such as the orbital paths of Earth and the Moon, also took Ebert as their model.

sum of more than 30,000 taler (cf. p. 982). So the utility of this animal was its financial value.

The treatment of the "whale-fish" provides a similar example. Following some introductory descriptions about the usefulness of fish in general, the viewer learns that whales are hunted with harpoons from a boat. The whaler's throwing gesture dominates the picture's message, however, not the whale. The illustration below this whaling episode is an even more blatant expression of the human prerogative over animals. The image devoted to herrings does not show the fish, but the fishery, complete with nets, fishing vessels, and dinghies. Any interest in the shape and appearance of herring must give way to its exploitation as a food source.

The second natural history work cited by Stoy was by the pedagogue and author of youth literature Georg Christian Raff, a private lecturer and deputy headmaster *(Konrektor)* at the lyceum in Göttingen, who since 1775 developed a teaching method that was remarkable for its dialog form (see Brüggemann and Ewers 1982–91, 1:1238). Raff's "Natural History for Children" *(Naturgeschichte für Kinder)* appeared in 1778 and comprises one volume of text with fifteen copperplates. It was a textbook for "children of every kind, rich and poor" (foreword) five years of age or older, and appropriate only for school use, to be read under the teacher's guidance. Raff portrayed the ethical aspect of the stories in the bold expressions of the animal characters and emphasized the "ennobling character" that such lessons about natural history might have for children. This pedagogue described the three natural kingdoms without going into physics, as Ebert did. Raff's illustrations are divided into three horizontal fields. The upper and lower ones, both narrower, depict the animals; the middle field provides many examples of their practical exploitability. How the animals might and must serve mankind was consequently given a central position here as well.

By contrast, Stoy did not just instrumentalize the animals for human ends but attached a second, ever-present field of significance: that of biblical Scripture. The whale was not just a good catch for humans, but also a safe haven for Jonah in its belly. The correlation between fish and fishing, on the one hand, and monstrous protector from biblical lore, on the other, reflects the Academy's central theme: man trusting in God, an animal world oriented toward and mediated by him. Stoy was not interested in the classificatory aspect of modern natural history (Linnaeus, Buffon), but in the practical utilization of whale and herring as food sources.

Ebert's and Raff's were the most well known works on natural history for children during the second half of the eighteenth century, besides those by

Abraham Trembley (1775) and Anton Friedrich Büsching (1775). Excerpted natural history was provided by Adelung and Weiße in their periodicals cited by Stoy. Their few illustrations were not replicated, in full or in part, in the Academy, however. Even so, many of the same anecdotally colored reports about various animals reappeared—such as the amenability of the marmot to taming, what the root of its name means, and what the animal is good for. When passages look familiar—such as the description of the nightingale in *Der Kinderfreund* (1780–82, pt. 2, 67), which resembles the one in the Picture Academy (Stoy 1780–84, 460)—we can safely assume that Stoy adopted its essentials. Nevertheless, Ebert and Raff remain the major sources for Stoy's design of his Academy, and all these texts have a virtually identical underlying pattern. As God's creation, nature is well ordered, nothing in it is without use and purpose. Hence it constitutes a harmonious framework in which everything has an explanation, leaving no room for superstition. Man is at the head of all living things, nestled within the web of nature: "If one observed and examined things more closely, one should find that each thing is made and equipped to a certain end, that one is available for the sake of another, that especially all lifeless things and all vegetation exist for the sake of living beings, that they all exist primarily for the use and enjoyment of mankind, and that one man exists for the sake of another" (Büsching 1775, 45).[12]

The Field for the Vocational Trades

> Field no. 6 is devoted to the common vocational trades, arts, and professions. In addition to Basedow's, representations therein are taken from the Viennese *Schauplatz,* Halle von Justi's, and Sprengel's works; and so that the young have a sure guiding line through all the arts and crafts, three aspects are always kept in view in the explanation that follows: what, from what, and with what the artisan works; or else the work itself, the materials and tools. (Stoy 1779, 7)

For field 6 Stoy provided by far the longest list of sources, which are best characterized by the contemporary concept of the spectacle or *Schauplatz.* This popular concept in the eighteenth century was "that which strikes the senses" (*ce qui frappe les sens,* Pluche 1733–39, 1:ix). That which impinges upon the

12. Although Stoy did not explicitly refer to the *Physica sacra* or *Kupfer-Bibel* by Johann Jakob Scheuchzer, it should be mentioned that the design and conception of the Nuremberger's book is in the same physicotheological tradition as this Swiss scientist's and hence in the tradition of correlating natural history with biblical Scripture (see Müsch 2000, 89–155). See also in the following section about the vocational trades the books by Noël Antoine Pluche; in this regard and for a general physicotheological background to such children's literature, see Philipp 1957, 23–31.

senses contrasts with that which is imperceptible to them and hence gains access to the mind only through reason. Old languages like Latin or Hebrew, scholastic theology, and philosophy are among such nonsensory objects of study and were the basis of all learning since the Middle Ages. Plants, animals, or physical principles, however, require a sensible, thus a visual or tactile, means of knowledge acquisition. These were rarely recognized as adequate subjects of learning, however, even in the eighteenth century. This concept of the *Schauplatz* mirrors the struggle for the establishment of realistic or practical instruction. Before this, scholarly training required no practical knowledge. Yet in the seventeenth century new economic demands arose in shipping, mining, and metallurgy, calling for a rethinking about what was useful knowledge. Mineral structure, the purpose of a mining shaft, and the composition of iron are fundamentals that the mercantile *Bürgertum* increasingly had to master. Only he who acquired such "reasonable" *(vernünftige)* knowledge could become a useful member of society. This is the context in which this children's literature emerged. Whatever reached the senses was not only more easily learned but also accessible to all men; for what ought to be the subject of study was the "Book [of Nature], which is so rarely read, even though it lies before every one's eyes" (p. v).

In his description Stoy first cited the work by Johann Bernhard Basedow, the *Elementarwerk.* The engravings from this book appear mainly in the fourth field and in that of the vocational trades (with a few also appearing in other fields in the upper and lower rows). The depiction of the painter's trade, for instance, appeared in Basedow's work, drawn by the artist Daniel N. Chodowiecki, who had taken as a model the same motif in Comenius's *Orbis pictus.*

The second source Stoy identified as a "Viennese *Schauplatz*"; more than one work could meet this description. The encyclopedic textbook *Schauplatz der Natur* by the French theologian Noël Antoine Pluche was first translated into German in Vienna between 1746 and 1753. The original French edition, *Le Spectacle de la nature,* was published in Paris from 1732 to 1739. It contains considerably more engravings, and in places the text is also longer than the German translation. Written for young nobles, this textbook discusses in conversational form the natural kingdom, the human world, and religion—whereby the doctrine of Christian salvation was taken as a basis—and thereby was distinct from natural philosophy. The work follows the tradition of edification literature for children of the upper classes, and a major portion of the discussion treats affairs of state. The copperplate engravings are basically stylized animals, plants, maps, and instruments without any "natural" background setting. The volume on professions depicts people at work, with their implements in the foreground. Two copperplates from the vegetable kingdom reappear in Stoy's work—in the field for natural history: the twenty-fourth engraving of seeds and the twenty-fifth plate depicting the inner parts of plants (Pluche 1760–66, 1:474, 481; cf. Stoy 1780–84, copperplate 42, field 5).

Another *Schauplatz* appeared with the collaboration of Johann Peter Voit between 1774 and 1783 in Vienna. Its title referred to a generously illustrated encyclopedic textbook on natural history and the arts in four languages with a strong orientation toward practice. Voit aimed to link language instruction with the other subjects and cited the *Orbis pictus* as his model. Each volume of this work, which appeared as a weekly serial published over a period of a decade, contains forty-eight copperplate engravings. Both the text and the plates catered to the taste of the *Bürger* and could be used in schools in general as well as in private instruction. Basedow's *Elementarwerk,* which likewise appeared in 1774, served as a model for these authors as well.

Stoy's designation "Hallens von Justis" may refer to more than one work, since the orthography here is quite confusing. Either it is a single work that was coauthored by Halle and von Justi, or else a comma is missing after the genitive of Halle, which then would not exclude the possibility that separate names and hence separate works are involved. Johann Samuel Halle, professor of *Staatshistorie* in Berlin, became renowned for his book, "Workplace of the Modern-Day Arts, or the New History of the Arts" (*Werkstäte der heutigen Künste, oder die neue Kunsthistorie,* 1761–79). Whether Stoy used this work or the *Schauplatz der Künste und Handwerke* (1762–1805) edited by Johann Heinrich Gottlob von Justi, on which Halle also collaborated from the fifth volume onward—thus, a work carrying both authors' names does exist—cannot be firmly established.

Halle's *Werkstäte* consists of six volumes of descriptive articles on either a single art or craft or a number of them together. Preceding each article is an engraved illustration of the relevant profession. At the end of a chapter are one or more foldout plates on which the instruments, apparatus, and tools for the manufacturing process of each trade are displayed. Each volume terminates with an alphabetical index. We encounter illustrations of the carpenter's or the apothecary's trade, which Stoy incorporated into his canon of pictures, albeit in a smaller rendition. They are slightly altered in design as well: in the Academy, depicted people look larger and are brought more to the fore.

Johann Heinrich Gottlob von Justi's *Schauplatz,* on the other hand, is a monumental work of twenty-one volumes. This cameralist used a work by the Parisian Academy of Sciences as his basis, translating the text with only slight changes and additions. It was supposed to serve "not only to further human knowledge, but also to be of use to human society, and especially to the country that nurtures it" (Justi 1762–1805, 1:8). The individual chapters are clearly structured. First the individual trade and manufacturing processes are introduced in a descriptive section, then a plate displays people at work and the tools and instruments used. Each plate is explained, and a glossary of special expressions related to the topic of the chapter follows at the end. Starting with the fourteenth volume, this work with its innumerable plates was published in Nuremberg. A study of sample texts and illustrations did not reveal any particular borrowings on Stoy's part.

Finally, Stoy cited the encyclopedic work by another teacher and pastor, Peter Nathan Sprengel had been teaching at the Royal Realschule in Berlin since 1762 until he left his position and became ordained in 1768. While still at the *Realschule,* Sprengel started work on his multivolume collection of "Handicrafts in Tables" (*Handwerke in Tabellen,* 1767–95).[13] The first fifteen collections had appeared by 1777 and bear Sprengel's signature. The work was intended as a handbook for young people from the middle classes, with which they could prepare themselves for their "future mode of living." Sprengel proceeded using a "tabular teaching method." He assigned, within a preset division of regularly recurring textual sections composed of an equal number of intermediary steps, the following content: materials, tools, products, and future prospects. Each section or career is illustrated with a folding engraving of the relevant tools. This is a much more thorough treatment than Stoy could afford to do with his Picture Academy, and a brief comparison yielded no direct borrowings here either.

As was the case with the field on natural history, Stoy used the most well known and recent literature of the time for this field on vocational trades. It harks back to the tradition of *Ständebücher,* books depicting a representative member of each of the professions or social stations.[14] Here too emphasis is placed on persons and their social settings—both in the textual description and in the illustrations. The manufacturing process and the ancillary materials or tools for a particular product do not play any significant role in the Academy. It is therefore not surprising that, unlike Sprengel and Justi, Stoy did not discuss factories or manufacturing plants: there is no mention of lead, calico, or gun

13. The final volumes, appearing in 1794 and 1795, were authored by O. L. Hartwig.

14. This connection was first made by Reuter. The *Ständebücher* relevant as sources for the Picture Academy are those by Jan Luyken from 1694 and by Christoph Weigel from 1698. The latter was embellished with aphorisms in an edition by the preacher Abraham a Sancta Clara. It was read as improving literature, in addition to illustrating the professions and social stations (see Reuter 1994, 98–100).

factories, for instance. Instead it is the gun maker and the weaver who make an appearance.

In his second volume, Stoy placed scholars at the summit of the pyramid of careers. Then follow the mechanical arts; the crafts associated with the making of clothes, houses, food; and finally, day laborers. All the vocations illustrated in the sixth field represent these two professional stations in the intermediary range of the social hierarchy. As in the field of natural history, Stoy made an effort to convey the interdependence of all careers of the middle classes: "All in all, one person is always indispensable to another; and many human hands must be occupied just to meet our barest necessities" (1780–84, 51). So a youth would "profit well by these practices, strengthening their efficacy and physical skills" (p. 146). The Academy extended a boy's or young man's horizons, while assigning him a central position among the orders of animals, plants, and human society: his own actions could contribute toward the nation's prosperity. However, the brevity and variety of Stoy's descriptions do not allow for completeness in the essentials for children and adolescents. Instead Stoy left open the possibility of looking up a key word and researching it further in a more specialized work.[15] Stoy's summaries and condensations are primarily conceived as preparatory material to put the reader or picture viewer in a position to learn more and to adapt and supplement the knowledge for his own purposes. But the fifth and sixth fields do reveal how very succinct the Academy is and how strongly it is oriented toward moral members of the working world. It corresponds to the utilitarian theory of professional training within a given social station (see Blankertz 1981).

The Central Field for Holy Scripture

The comprehensive layout of Stoy's Picture Academy makes it an elementary work; the universal quality of its index makes it encyclopedic. Yet unlike an encyclopedia, Stoy's work is not arranged alphabetically. The system in accordance with which he organized his material is biblical history. The correlations between the fifty-two individual copperplates of the volume were drawn from the Creation story in the Old Testament through to the New Testament's revelations. Stoy chose to use the Bible as a basis for extending common knowledge because it was already firmly established in the minds of child and parent alike.

15. The brevity is particularly obvious where a number of careers are summed up in a single article (see Stoy 1780–84, 641, the article on weavers and tailors).

> From among the biblical stories and events, only 52 have been chosen, two-thirds from the Old Testament and one-third from the New, so that young people can easily survey them and grasp them within one year. Yet for most of the main illustrations from the Holy Scriptures, the most closely related ones are added in smaller divisions, some of which, e.g., for the miracles of Moses and Christ, will number 10 or 12. (Stoy 1779, 5)

This is how Stoy described the central segment of the Picture Academy. Unlike the other subjects, here he dispensed with citing any sources and seems to draw from common, generally held knowledge obviating further comment. This is not surprising, considering that picture Bibles, which first started to appear at the beginning of the seventeenth century, were particularly popular by the turn of that century (see Bauer 1982, 887). The lack of citations also supports the assumption that, although giving Holy Scripture a central place in the Academy, Stoy assigned it primarily a mnemonic function, grounded in children's existing, though incomplete, familiarity with the material, to form a point of reference for all the other fields. The Picture Academy thus emulates what are known as history Bibles, which summarize general and biblical facts, closely linking faith with learning. The result is a form of illustrated Bible-oriented history of mankind.

In discussing the content of this pivotally important book, Stoy resorted to pigeonholing diverse anecdotes rather than reporting on its morally or theologically significant points. Where it was not possible to devote a separate plate to each of Christ's miracles, they were boldly cut into two classes, making the plots proceed not chronologically but by a series of climaxes.[16] In the illustration, a central, important event is depicted in the middle of the field, and the other incidents are arranged around this center. In this way Stoy provided structure and orientation among the abundant—if not overabundant—individual biblical scenes. A child might initially divide them mentally into two parts, according to their origins (the Old and New Testaments). The novel, abbreviated arrangement depicts in a simple manner actions by people whose claim to fame lies in their appearance in the annals of biblical history. The point here is not communication of the faith but remembrance of individual deeds, to which other deeds in the adjacent fields can be linked. The arrangement of the fields mirrors the Christian point of view that all areas of life and all knowledge are based upon a rational order made and fully controlled by God. A more worldly interpretation—and the one that is prompted by the boxed

16. The same device is used for the ten plagues of Egypt (copperplate 14), the passion of Christ (copperplate 48), the apostles (copperplate 50), and various Christian martyrs (copperplate 51).

version of the Academy—sees the Scriptures simply as a basic reference from which to point to things of relevance to a child's everyday life. Bible stories could be used in this way only because they were part of a common, centuries-old knowledge, particularly among the lower social classes and stations.

Catechisms as tools of religious instruction first appeared in Luther's time.[17] They were illustrated from the very beginning with associated commentary. Initially, simple, straightforward woodcuts were used to illustrate the verbal message. Since the sixteenth century, and particularly in the succeeding baroque period, an emblematic element was added to these biblical depictions. Such images contain many interrelated levels of meaning, transforming the image into a symbol that carries substantially more potential associations than might at first be assumed. Stoy's Academy was at the end of this development of emblematic depiction, which increasingly lost ground to a more "realistic" one in the second half of the eighteenth century. Both trends meet here: the associative, ambiguous, and symbolic representation and the "realistic" one, with motifs borrowed as much from recent knowledge about natural history as from a child's environment.[18] In the text the change manifests itself in the choice of a style more suitable to children and the addition of different explanations for fascinating, striking facts.

Most of the illustrations in the central field are more or less identifiable excerpts from existing biblical works from the seventeenth and eighteenth centuries. Stoy was not the only author to use the copperplate engravings by Matthäus Merian the Elder as a model. Between 1625 and 1627 Merian developed a series of etchings covering the entire Bible, which were subse-

---

17. Luther defined a catechism as "a lesson for heathens wishing to become Christians, so as to teach and guide them in what they should believe, do, relinquish, and know" (cited in Fraas 1988, 710). Until the nineteenth century, catechisms, along with edification literature, fulfilled not only the basic function of strengthening the faith of believers but also of providing subject matter for lessons in reading and writing. The catechism is the lowest common denominator of Christian or Protestant doctrine and is one of the most effective supports in the church's mission.

18. Since the sources of the pictures in the upper and lower rows are not individually described but only generally discussed in the following section, let me demonstrate the "modernization" of illustrative expression with one example of a picture from the mythology field contrasted with an engraving from Sandrart's *Ovidii Nasonis*. In this scene of the creation of man, the background landscape is originally European (cf. Sandrart, undated, copperplate 2); the Academy's rendition of this same motif embellishes the European flora and fauna with an elephant, antelopes, and a bird of paradise (Stoy 1780–84, copperplate 2, field 8). Although a parrot does make an appearance in Sandrart's work (fulfilling an emblematic function), in the Academy more recent natural history is incorporated into the design. In addition, Stoy has his "new human" modestly cover his private parts with his hands, while Sandrart depicts him in unabashed innocence. Thus the copy is converted into a form meant to satisfy children's interests while conforming with the current standards of decency.

quently recopied countless times. These pictures were first bound into Luther's translation of the Bible in 1630. Stoy's illustrations of the wise judgment of Solomon in mediating between the two quarreling women and the construction of a temple in God's honor are both based on Merian.[19] The king's pose, the picture composition, and incorporation of the narrational motif within the setting are all virtually identical—as far as can be made out in the reduced copy of the Academy. The first impression each gives is interesting, though: the temple complex seems considerably farther away in Stoy's version. In the original, the neighboring houses and villages are nestled right up against its peripheral walls, whereas in the Academy, a broad strip of open space separates the complex, viewed from a central perspective, from the surrounding area. The vanishing point clearly lies in the almost absolutist reign of the "great church, which was called the hallowed house" (Stoy 1780–84, 472).

"In the 17th century . . . engraved and etched Bible series attain wide circulation and great popularity, whereby the decades around 1700 are particularly productive" (Bauer 1982, 887). The southern German area surrounding Augsburg and Nuremberg and the regional hub of engraving and Bible illustration in the seventeenth century could boast of Merian as well as the engraver, sketcher, and art dealer Christoph Weigel. Stoy must have been familiar with his work, since Weigel had built up a large art dealership and publishing house in Nuremberg at the beginning of the eighteenth century. Among the first picture Bibles that he designed, in collaboration with the artists George Christoph Eimmart the Younger and Johann Jacob von Sandrart, was the *Biblia Ectypa* from 1695.[20] After the great success of the *Biblia Ectypa*—also financially speaking—Weigel embarked upon a similar large project in 1705. This illustrated Bible eventually appeared in 1708 under the title *Historiae celebriores veteris testamenti iconibus repraesentatae.* As with the *Ectypa* edition, two other famous collaborators were also engaged in this work, the Dutch artists Jan and Caspar Luyken.

The most direct conclusions about the original models for the engravings in the Picture Academy can be drawn from Johann Rudolf Schellenberg's illustrated Bibles from 1779. Since he collaborated on Stoy's work and essentially drew up the New Testament pages, we may take one illustration (figures 6 and 7) to represent other borrowings and reproductions from it. In reworking the

19. Cf. Merian 1630, 179, 187; Stoy 1780–84, copperplate 26, field 1.

20. Bauer ascribes this coinage to Weigel and surmises that he used the various Greek meanings "to present his Bible simultaneously as a pictorial narrative and as a model collection of illustrations" (1982, 887).

Figure 6. Jesus and the children in Schellenberg's picture Bible (1779)

Figure 7. Jesus and the children in Stoy's Picture Academy (1780–84, copperplate 41, field 1)

Figure 8. Job in Merian's illustration (n.d. [seventeenth century], pt. 2, plate 77)

subject of Jesus and the children, the artist retained the principal actors of the original print and altered only the foreground and background.[21]

In many places the Picture Academy indicated the opposite tendencies from those in its predecessors, however. One example is the story of Job, a rich man deeply devoted to his Lord and Creator, whose resolve and unshakeable faith is put to the test: besides suffering great pain, Job must also endure harassment by his wife and relatives, who try to convince him that God wants to punish him. This moment, when Job appears not to believe them but still feels the intense pressure exerted on him, is the subject of one of the engravings in the illustrated Bible by Matthäus Merian (figure 8). Calm and patient Job is sitting in the foreground of a dramatic scene, facing his wife's verbal assault. A clawed demon with female breasts is hovering above his head, but even this apparition does not seem to unsettle him.

21. Reuter cites in addition to the *Biblia Ectypa* the *Poetischer Bilderschatz* of 1758 (which reproduced engravings from Weigel's *Biblia*) as another model for the Academy (1994, 82).

The Picture Academy, by contrast, omitted this incarnation of Satan from its version of the beset Job (figure 9). Sitting in the midst of his arguing relatives and confronted by his scolding wife, this Job looks visibly perplexed (see also the accompanying text, Stoy 1780–84, 651–55). In this picture, the battle against traditional conceptions about evil spirits has already been won, and the fanciful baroque illustration has developed into a more rational and sober motif focusing on human interaction.

The depiction of the sacrifice of Isaac is another example. At God's command, Abraham is about to sacrifice his son as a burnt offering. Having bound and laid Isaac on the prepared altar, he is poised to slay him when an angel appears in the heavens and puts an end to this test of his faith. This dramatic instant when the angel comes to save Isaac's life is a popular biblical motif and a recurrent pictorial theme (see Kutra 1906). Johann Hübner's picture Bible (1759), with copperplate engravings by Peter Conrad Monath from Nuremberg, has the angel appear in the clouds to interrupt the sacrifice. Monath, for his part, was inspired by the illustration in Weigel's *Biblia Ectypa* of 1695 (plate 5). Basedow's rendition of it in his *Elementarwerk* of 1774 takes a more enlightened approach to religion, based on a more modern state of knowledge, and leaves the angel out of the picture (plate 6). If one follows Abraham's

Figure 9. Job in Stoy's Picture Academy (1780–84, copperplate 34, field 1)

line of sight, though, the angel's contours can still be made out in the sky, the remnants of a feature that was erased from the copperplate.[22] Thus Basedow did away with the angel, and even the text is evasive: "But as he was about to drive the fatal steel home, he was stayed (according to this version) by a new revelation cautioning him not to complete the act, and his ready obedience was praised" (1909, 2:232). Stoy, however, reinstated the angel in his illustration of this episode (plate 7). Here the angel is even grasping Abraham's arm to stop him from killing his son: "Now Abraham took up the knife to slaughter his son, who patiently let all this come to pass. But lo! an angel called out to him from heaven: Abraham! Abraham! Here am I, said Abraham. The angel, then: Do no harm to the boy! God does not desire his blood" (Stoy 1780–84, 137). Both verbally and visually, this scene is depicted in the Academy with considerably more emotion and drama. Basedow's distance has been completely retracted.

These examples show how difficult it is to make any generalizing statement about the Picture Academy; more often than not, the contradictions or lack of clarity confuse the viewer or reader.

The Fields in the Upper and Lower Rows

Whereas the fifth and sixth fields contain factual information, the fields in the upper and lower rows convey moral messages. Such fables and myths were compulsory topics in eighteenth-century education: "one must know the Fables in order to understand the works of ancient and modern culture" along with their corresponding "ornamental accessories" (Starobinski 1989, 234), like certain characters, symbols, and modes of expression. Stoy emphasized the lessons of comportment attached to the moral meanings of the depictions. Only in rare cases are exemplary tales, anecdotes, myths, and legends presented as masterpieces of ancient cultures.[23] Guidance on good behavior is consistently distilled out of them. They mirror the omnipotence of God or the transience of humanity, and are full of recipes for life, admonitions to use forethought, and instructions about correct conduct. These fields contain a variety of themes: mythological topics are not restricted to the eighth field but are also found in the third and seventh fields. The field for "older and more recent world history" does not limit itself to its designated subject, either. Fables appear in contexts other than the seventh field, and allegories are scattered throughout the work. These rows serve as a collecting pool for bits of knowledge not accommodated in the middle row.

---

22. This point is made by Ringshausen (1976, 107).

23. For details about the inspiration that the Academy, among others, received from works by Bernard de Montfaucon and Joachim von Sandrart, see Reuter 1994, 106 f.

If Stoy's method is interpreted as *typological* borrowing, one may conclude that the upper and lower rows symbolically carry out the messages in the middle axis. It is not the "objects" *(Dinge)* that are adopted from the middle fields, but the associated rules of conduct. The fifth, sixth, and first fields carry the realia of the Scriptures, whereas the other fields carry the "ideals" in the sense of moral and spiritual guidance. For example, the biblical field in copperplate 12 relates the story of Joseph as the ruler who saves his land from famine by building large granaries during the years of plenty. The fields of natural history and vocations pick up the theme of grain and the milling and baking trades. The remaining fields discuss the duties of vassals, a ruler's patriarchal feelings toward his people, and the insignia of power.

Another distinguishing factor between the middle row and the upper and lower fields is the formal representation. The fifth and sixth fields draw their visual representations from encyclopedic literature; hence these are depictions of objects or occupations. The other fields tap a much broader-ranging tradition of emblems (in this case also biblical ones). "It is not clarification of the mind that we expect of the arts, but practical reminders of entirely familiar but very useful truths; not new concepts, but daily and vivid reminders of the most important and already well enough known concepts." That is how Johann Georg Sulzer defined the importance of emblems in his article on the *Sinnbild* (*Allgemeine Theorie der Schönen Künste,* 1771–74, 2:1081). Reducing the full scope of visual and functional forms to a single definition, an emblem links image with text and offers a learnable canon of representational forms.[24] Such an image always carries with it an appeal, a call on human attitudes.

Stoy also used this emblematic form in his Academy. The sources of some motifs can be reconstructed and thereby reveal the artistic elaboration of the motifs. The emblem of the honey thief from a Dutch emblem book of 1633 (Henkel and Schöne 1967, lxi, 921) explains that many painful bee stings are proper punishment for a greedy sweet tooth (figure 10). In Stoy's depiction, a boy who is trying to pick a honeycomb out of a hive is standing amid the bee swarm, suffering the same fate as his seventeenth-century brother in spirit

24. Emblems consist of an icon *(pictura),* taken from the most disparate sources going back to antiquity and bearing within them an inscriptive tradition many centuries old. Then there is the *inscriptio,* a kind of motto inscribed above. Below the *pictura* is the interpretive and explanatory *subscriptio,* which offers words of wisdom drawn from the figure. "Collectively, an emblem's inscriptio, pictura, and subscriptio perform the dual function of representation and interpretation, depiction and explanation" (Henkel and Schöne 1967, xii). It is only by means of the legend below that it is possible to make a connection between the motto and the figure. Warncke (1987, 161–92) criticizes the three-step definition of emblem described above, however. He argues that the figure and its explanatory legend are the fundamental characteristics of an emblem (pp. 165 f.)

Figure 10. Emblem of the honey thief (1633) (Henkel and Schöne 1967, 921)

(figure 11). Had he followed his father's advice, the excruciating stings could have been avoided. By incorporating emblems into an obvious educational process, Stoy modernized this traditional representation.

It was not always possible, however, to place such emblems within a new, more contemporary context. Stoy had some difficulty in making his purpose as clear as that of his predecessors:[25] At the highpoint of the art of emblems, in the sixteenth and seventeenth centuries, viewers learned to interpret the emblematic forms and to understand them even out of context (Henkel and Schöne 1967, xix). But in the eighteenth century this was not taken for granted. Stoy tried to translate this old teaching instrument, adopting the basic design, but altering the content or its message to conform better to a child's mind or to his personal moralistic point of view. Emblems operate on the assumption that the world and its phenomena were "replete with secret indications and hidden meanings, with veiled, hence discoverable signs" (p. xv). Stoy, however, did not seek out these inner meanings. Instead he applied a representational

25. This is the case for the depictions of Deucalion and Pyrrha as well as of the luminous yet self-devouring candlestick, which are taken from the emblematic tradition of the sixteenth and seventeenth centuries (see copperplate 5, field 8, and copperplate 8, field 2).

Figure 11. Depiction of the honey thief in Stoy's Picture Academy (1780–84, copperplate 3, field 4)

form that postulates the relation between things only from a formal aspect. He employed a handed-down method to draw relations between external signs or signatures (see the story of Joseph, related to grain and the baker's trade). The connections seem to be arbitrary, and the network of meanings seems not to hold anymore within the given order (see also Warncke 1987, 76). This is the reason, for its contemporary reviewers, that the Academy's order seems contrived rather than sensible.

Copperplate 38: "First Class of the Miracles of Christ"

If we consider the various physical forms in which the Academy may be used—book, poster, and box—the box is fundamentally different in that it permits separate viewing of the picture cards of a single field and therefore of a particular topic. In tableau form, by contrast, the individual picture fields are always perceived in relation to their immediate neighbors. The separate picture segments must be read as supplementing one another if any sense is to be made of their arrangement. Consequently, an interpretative dimension is added to the information provided by the text. In the following I discuss a single copperplate, to demonstrate this mutual influence among the fields.

Stoy treated the miracles of Christ on copperplate 38 in just the same way that he handled fish and birds: he divided them into classes. The works "with which He revealed Himself as Lord over all things, can be ordered into two main classes. To the first belong all the miraculous cures of the Savior, whereby He healed those afflicted in body and mind; and to the second, all the other wondrous events, transformations, and alterations of things" (1780–84, 768). The first main class, Christ's miraculous cures of the sick, is the subject of this copperplate (plate 8). The healing episodes depicted here are occasions for Stoy to report certain surgical operations on eyes and ears, as demonstrated in the fourth field. The anatomical structures of these two sensory organs appear in the fifth field. The eighth field presents Aesculapius, the god of medicine, and the seventh field, the two deities Minerva and Apollo, pitting their powers of healing against each other. The remarkable accomplishments of a blind woman (field 9) show that it is possible to learn to move about even without that most important sense, vision. The unrelated fields 3 (from the Islamic tradition) and 2 (electricity) are unceremoniously added to this plate about mastery over illness.

The relations between the individual fields were introduced above. We saw that some correlations were too vague, others forced. This copperplate suffers from the same problems. Some themes adhere to the Academy's intention of deriving everything from the central field and adopt elements directly from Bible stories. Other connections are laboriously made, and two apparently have nothing of substance in common with the others. Taken alone, the copperplate's individual fields are very detailed, even minuscule at some places (plate 9; field 1 is composed of eleven segments), and intense study is necessary to make everything out. The engravings on this copperplate and of the Academy as a whole show the effort to achieve a more realistic representation, in the sense that a more everyday look is given to what is being depicted. The Academy altered the material from older sources and updated the drawings and their composition. This might take the form of moving a person more into the foreground, blotting out an angel, or depicting the woods in more than an outline. These illustrations are entirely the work of a new, enlightened, and complex view of the world.

The second field of copperplate 38 carries the heading "Physical Experiments." After a brief and general introductory passage, Stoy starts with a description of electricity (plate 10). The depicted experiment is conducted with an "electrostatic machine" (1780–84, 772). A man is standing beside this machine, touching a rotating glass sphere with his hands. The electric charge generated by friction is conducted by a wire. "This wire is touched by a person standing on resin, who sets fire to spirit of wine, which is held out to him in a spoon, with his extended finger" (p. 773). While Johann Jacob Ebert referred to physical experiments as "practical instruments" *(nützliche Werkzeuge)* in his *Naturlehre* (1776–78, 1:176), Stoy described them as the "finest experiments" (*schönste*

Versuche, 1780–84, 818), which he saw as proof of God's almighty power. While Ebert incorporated a description of Guericke's air pump in the letters "Explanation of the Air and the Air Pump," "About the Elasticity of the Air," and "About the Gravity of the Air and of Airships," Stoy lumped the air pump with the burning mirror, Magdeburg hemispheres, speaking trumpet, barometer, magnet, and Cartesian devil (see copperplate 39).

This is not particularly surprising, considering that Stoy relegated experiments, stripped of their contexts, to the realm of popular curiosities, or at least wrote them off as not entirely explicable. The eighteenth-century experiment, especially the spectacular phenomenon of electricity, was a social occasion or a children's show: impressive, yet not particularly complicated (cf. Stoy 1780–84, 230, 774).[26] This curiosity element is implied in the Academy's commentary on the electrostatic generator, but the illustration does not provide any more details. It suits the general pattern of presenting this experiment. The fourth and sixth fields draw the eye to well-lighted rooms. Operations or a medical examination are all being performed by men in appropriate attire and attitude (plate 11). The rooms are simply and practically furnished. Nothing superfluous distracts the viewer's gaze, and he can watch while taking a glimpse behind the scenes into the world of medicine.

In the central field, miraculous biblical cures are taking place. Jesus is healing the lame *(b),* the blind *(a),* the deaf, and the dumb (*d,* plate 9). In art, Christ's wondrous works always had "far more significance than the other scenes of his influence" (Schiller 1966–91, 1:171). Although very different symbolic images have arisen, these miracles are always directed at human beings. "Curing the blind is one of the most frequently illustrated miracles. . . . Opening the eyes means being led from darkness into the light; in the metaphorical sense, from lack of faith to faith" (p. 179). Christ touches a person or points at him with outstretched arm and cures him by the power of his divine word. In healing the blind, he touches the blind person who is kneeling or standing before him, covering the sightless eyes with spittle. A

26. It is surprising that Stoy restored the full force of the element of mystery after Ebert had already incorporated such instruments within a rational framework. While Ebert only described hitherto unsuccessful attempts to construct an airship, Stoy provided an illustration; while Ebert depicted two air pumps so as to illustrate their differences and to give a better idea of how the apparatus works, Stoy exhibited a single unit and, directly below it, the Magdeburg hemispheres. Anything with a scientific flavor, such as a description of an experimental setup and its associated experiment, Stoy pulled out of context and depicted as isolated instruments, in order to transmit the necessary knowledge about the current state of natural philosophy, especially popular natural philosophy. About the popular experiment spectacle of the eighteenth century, see Schaffer 1983 and Sibum 1990.

counterpointing arm is one of the earliest symbols of this miracle. It is of particular importance in copperplate 38 and shows that the Academy's picture tableau conveys an entirely different message from its associated text, for medical operations, the physician's career, and physical experiments have nothing in common with miracles—at best, curing body or soul. Although the contents of the individual fields are somehow interrelated, they are not equivalent. Stoy's claim in the text that experiments were artful is not sufficient to equate them with miracles. On the formal level of illustration, however, Christ's motions are at the focal point of the central field: hand and arm positions indicate Christ the healer, the blesser, the actor. We see these miracle-making gestures replicated in the other peripheral fields. The same extended arm touching the spoonful of spirit of wine (electricity) or the blind man's eyes (miraculous cures) implies a similarity between the actions in a way that the descriptive text alone could not. The outreaching arm appears in the fourth field (plate 8). of surgical operations, in the sixth with the bedside doctor, and in the seventh with administering Apollo. The first impression conveyed to a viewer still ignorant of the passage accompanying this sheet is of a haphazard conglomerate of tiny engraved images of building interiors and mountainside or city scenes. It is only upon closer examination that the unifying gestural element of the outward-reaching healing or blessing hand becomes apparent.

The purpose Stoy attached to this tableau becomes clear from his definition of a miracle. Wondrous works were "such deeds as no man is able to perform and do not obey the normal laws of nature" (1780–84, 768). Electricity and the physician's feats then follow, both of which, however, were rapidly losing their incalculable mystery in the eighteenth century and no longer counted as miracles. Viewed thematically from the perspective of the individual fields—in this case the medical career stands between the engraving of wells and fountains, and the apparatus and work of experimental physics (on copperplate 39)—such an association with the obscure and enigmatic does not arise, Stoy's few remarks about miracles notwithstanding. Stoy intended to represent scientific discoveries as heralds of a modern age, not unlike the church with its tidings about the Messiah's marvelous deeds in attestation of his divine power and as a sign of kingdom come. Moreover, the gestures establish a formal similitude and equality that no longer differentiates between symbolic interpretation and realistic illustration. Many of the tableaux have this representational mix, with the topics grouped according to formal aspects.

The illustration tradition of Christ's miracles must have had some impact as a conscious or subconscious model for the depiction of experiments in the eighteenth century. But Stoy went one step farther. With his picture tableaux, the eye cannot resist the impression of simultaneity among the assembled images and must necessarily draw connections between the individual fields in order to understand their layouts and deeper meanings. Depicted gesture provides the unifying formal element for copperplate 38 and acts as its basis. In this copperplate, textual material and representational form contradict each other.

One optional use of the Academy was as "a manual, a collection of selected material and illustrations, a store of good food for the minds and hearts of the young. Even adults will find much in this book that they either may have long since forgotten or may not have easily found summarized so succinctly, intelligibly, and accurately from several works" (Stoy 1780–84, directions, 1 f.). Even this positive characterization by Johann Siegmund Stoy indicates the problematical points of his Academy, indirectly stating its ambivalent features. The mentioned brevity of the teaching material and its forced alignment to Holy Scripture condemns the Picture Academy to a position inferior to the more comprehensive specialized works. Both text and copper engravings place man at the center. The Academy's purported focus on Christian doctrine and symbolism is used as the unifying rationale for its emphasis on human purposes, deeds, relationships, and faults; but thereupon it negates itself. Consequently, the Picture Academy cannot claim equal incorporation of spiritual and worldly knowledge. It is rather a "humanized textbook," a book for and about people. It chooses mankind as its topic and presents it in easily digestible form.

Contradictions between Christian and secular content recur among the pictures (illustrations with superstitious elements removed—as in the story of Job—and reinstated again—as in the story of Abraham) and between picture and text (scientific experiments and Christ's miraculous wonders). The effectiveness of the pictorial image itself is diluted by the minuteness and sheer abundance of detail. Stoy inserts labels in the cluttered illustrations as a means of orientation, or isolates individual depicted objects—with a consequent loss of simplicity. Hence Stoy's compendium of knowledge can scarcely provide a serious basis for thorough elementary education. It cannot provide adequate preparation for vocational training, nor can it sufficiently document any one of the subjects it lists.

Instead, the Picture Academy provides subject matter for conversation. It is a schoolbook of stimulating facts, in which moral issues are presented on both the religious and secular planes. Stoy was a rational theologian who aimed to provide a practical and realistic educational basis for faith or the Christian catechism. He was, moreover, an innovative "artisan of education" *(Erziehungskünstler).* In adopting the more modern epistemological element of using images as an educational means, he chose visualizable knowledge as the basic medium of his work.

Tableau A

Plate 1. Frontispiece for the Picture Academy (Stoy 1780–84)

Plate 2. Frontispiece for Voit's *Schauplatz* (1774–83, vol. 1)

Plate 3. Dedication plate for the Picture Academy (Stoy 1780–84)

Plate 4. Depiction of a classroom from Stoy's Academy
(1780–84, copperplate 36, field 4)

Plate 5. Abraham and Isaac in the *Biblia Ectypa* (Weigel 1695, 15)

Plate 6. Abraham and Isaac in the *Elementarwerk* (1774) (Basedow 1909, plate 79)

Plate 7. Abraham and Isaac in the Picture Academy (Stoy 1780–84, copperplate 9, field 1)

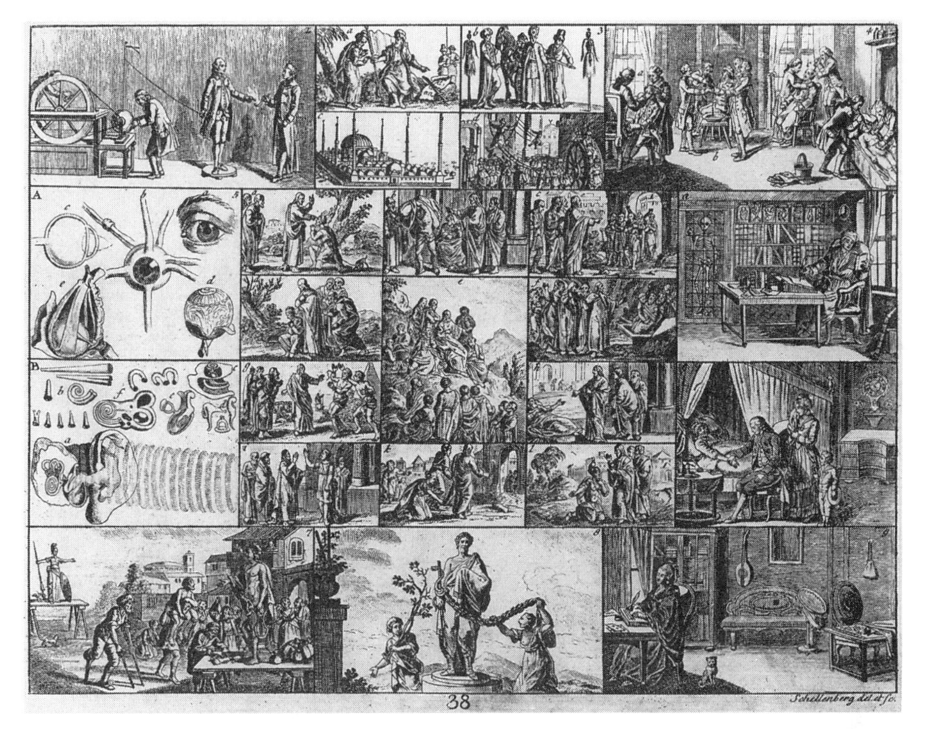

Plate 8. Copperplate 38 of the Picture Academy (Stoy 1780–84)

Plate 9. The miraculous cures of Christ (Stoy 1780–84, copperplate 38, field 1)

Plate 10. An electrostatic machine (Stoy 1780–84, copperplate 38, field 2)

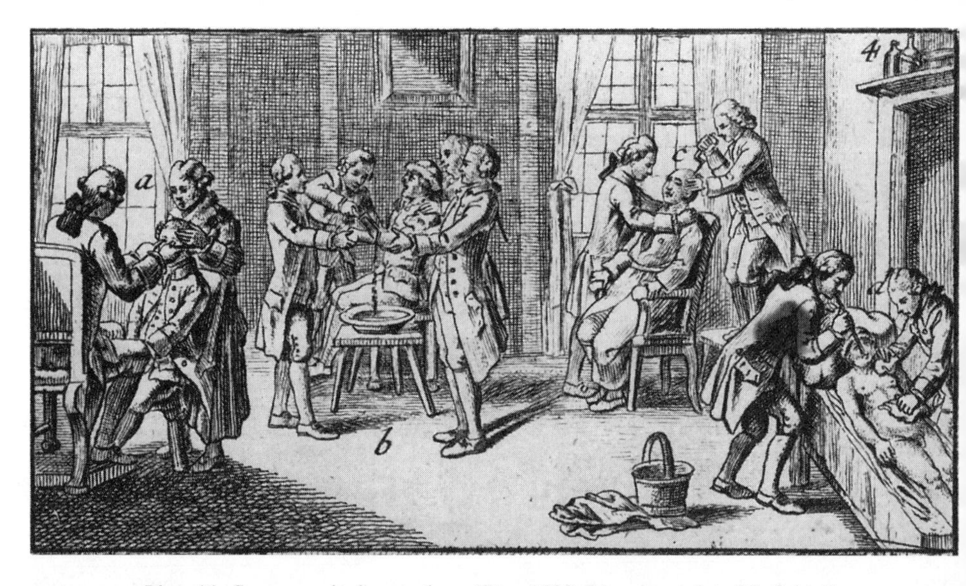

Plate 11. Some surgical operations (Stoy 1780–84, copperplate 38, field 4)

4

MAN AND HIS IMAGE IN THE EIGHTEENTH CENTURY

There were two ways of conceiving the human body in eighteenth-century pedagogy. On the one hand, it was a living, developing organism that had to be nurtured and guided through a specific phase in life. On the other hand, it was a figure depicted in educational books. Representation of the human body in illustrations—in copperplate engravings—was expected to impart guiding morals to children. What emerged was a wide range of changing associations connected to the body as a depiction, along with the physical effect of such depictions on the perceiving body. In this context Stoy's endeavor took part in a larger debate of the Enlightenment.

The Human Body

In the eighteenth century the human body featured as an object of discussion within the contexts of medicine, empirical psychology, and anthropology. Out of the science of man arose the concept of the consummate human being, the product of a carefully planned education incorporating the ideal of humanistically perfectible nature. The various disciplines play a part in this education: medicine, theology, and pedagogy, as well as philosophy and aesthetic theory. The Cartesian sundering of body from soul, dividing the person into a part metaphysical, part physiological being, was beginning to lose its sway in the eighteenth century. Ongoing controversy ranged from dietetics to the moods and passions, from empirical psychology to the dark, subconscious side of the personality. Education—the aspiration of bringing the human being to perfection—occupied a special place in these debates, as it was being institutionalized in the form of journals, books, and therapeutic facilities.[1] It involved

1. Wolfgang Riedel surveys the research on "literary anthropology" and its context since 1985. Although this study dates from 1992, his description of the various research topoi and his closing assessment of the literature in historical anthropology are very informative (see Riedel 1994; see also vol. 14, issue 1, of the journal *Das achtzehnte Jahrhundert* from 1990).

training a child's mind to acquire knowledge together with moral precepts. The body played a central role in the basic pedagogical ideas of this period. Education developed via the stimuli received by the senses, the child's perception of aesthetic concepts, and their effect on his or her body.

Attention to health and the responsibility an educator had toward the physical constitution of his young pupils was pointed out by the pedagogue Peter Villaume in 1787. He started by criticizing school instruction, which concentrated on the education of the soul but not of the body. "The body must be schooled, so that the soul may develop its powers thereby" (p. 233). As an implement of the soul, the body required special care and attention; it should be properly maintained and should become "robust" and "sound" (p. 286). Villaume placed much importance on eyesight. It was not enough that a child's "eye be bright and see acutely, that it see and distinguish tiny remote objects: it must be able to survey much at a single swift glance, so as to be able to decide rapidly and act accordingly" (p. 285).

Conviction that sensory experience had a positive effect on the development of a child's learning ability stemmed from the philosophy of sensationalism. In contradiction to the "innate ideas" of rationalism, John Locke saw the soul as an empty vessel or blank slate on which the external sense perceptions accumulated over a lifetime. Thus, as the stuff of thought and knowledge, the empirical world was ennobled to an unprecedented degree. In conjunction with inborn human attributes—intelligence, carnal desires, a will—"compounded ideas" were formed to make up the content of consciousness. The debate over the reciprocal conditions set by the physical, psychological, and moral elements of human nature continued throughout the eighteenth century. This is the context in which Locke's writings were received, along with those of Étienne Bonnot de Condillac, Charles Bonnet, and Gottfried Wilhelm Leibniz, whose positions are essential for understanding the Philanthropin school. What they have in common is the important place they concede to the body and its sense perceptions in a child's intellectual and rational development.[2] Methodologically, the Philanthropinists followed a program of inductive empiricism—which since Francis Bacon's time had been a central model for acquisition of knowledge—and it was observation that enabled educators to assess a child's progress.[3]

Man was to be the object of observation. Illness or health could be read

2. Christa Kersting differentiates further among the Philanthropinists in her study of the genesis of pedagogy in the eighteenth century (1992, 115–67). She traces education at the close of that century back to influences on philanthropy, particularly to its forerunners in philosophy and natural history.

3. The elements borrowed from this empirical method, mapping human knowledge by means of observation and experience, Menze (1966, 31 f.) sees as the onset of pedagogical theory or a new

off the body's surface; it was openly viewable as a topic of education and demonstration. At the same time man himself was supposed to perceive the surrounding environment, take note of deviations from the norm, and quickly grasp complicated phenomena. These demands—not simply to look at things, but to use each glimpse to observe and systematize—were exclusively tuned toward the question of how body and soul could lead to moral upbringing and what resources and instruments could be taken up in this endeavor: on the one hand, compass and yardstick (see Stafford 1991), along with stopwatch and the gymnast's horizontal bar (see Eichberg 1981); on the other hand, a picture tableau from a Nuremberger's children's book.

New Signs: Expression and Gesture

Over the course of the eighteenth century the body acquired a significance of its own for the *Bürger*.[4] New forms of presentation and association were found to express the social hierarchy. Special attention to superficial comportment, including gesture, expression *(Mimik),* and posture, enabled a *Bürger* to set himself clearly apart within the existing social framework. Any behavior falling short of these rigid standards was a sure sign of inferior rank. The external signs of the body could thus simply be read off for identification of a person's proper social standing. Gestures and mime immediately revealed to the "trained eye" *(geübter Blick)* information about the invisible soul of the object of such scrutiny; hence they served as a banner of moral constitution.

Physiognomy—a *physische semiotik* (Goethe), *Semiotik der Affecte* (Lichtenberg), or *Semiotica moralis* (Christian von Wolff)—manifested the "body's speechless language."[5] The body speaks on the basis of proportions and comparative measurements. A silhouette machine draws not only the profile but also the character of the person whose portrait is being taken. Physiognomy was thought to reflect a person's genius. Toward the end of the century head shapes were investigated according to the precepts of cranioscopy, which is based on the assumption that there is a correlation between the internal (and

---

phase of it. Kersting (1992, 195–228) puts this aspect into the sphere of medicine in her analogy between the arts of healing and of child rearing.

4. Here, a *Bürger* is not defined in the general sense of the legal citizen of a state but in his or her capacity as a member of a social station. Norbert Elias (1976) groups his reflections about the taming and rationalization of the populace under the rubric "process of civilization"; his representation has been developed into an influential, but hotly debated, thesis (for more recent criticism, see the summary by Roeck [1991, 101–8]). On "geometrization" of the human body in the Renaissance as an internalized form of society and hence a history of the early *Bürger,* see zur Lippe 1974.

5. That is how Böhme summarizes the outward impression of what signals "the secrets in the hearts of other people" (1989, 148).

invisible) faculties of the brain and the external form of the skull. Accordingly, these external traits provided clues to the inner life, character, and moral attitudes of the person being subjected to a tactile inquiry. To this deciphering of the corporeal alphabet, primarily from the shape of the head, were added gesture and posture, whereby the whole person presented himself with every articulation of arm, foot, or hip.

The development of bodily signs indicates a shift of interest toward the individual or, as Porter puts it, the replacement of a "divine frame of reference with a human" (1990, 72). With the bond to theology still unbroken, an anthropometric catalog of characteristic traits—a classification of body language that everyone might understand—was compiled. This preoccupation with one's inner self sprang from the wish to penetrate to the ultimate truth about oneself and with the aid of the probing, deciphering eye to assign to the invisible the same discernible status that the visible already has. Anthropometry was based on a network of connections among philosophical aspects, scientific findings, and techniques of visualization. Its purpose was to find the natural language of man. Such insights also allowed a higher degree of awareness about human manipulability and control, particularly over children. Control over the body is exerted not just from the outside; education is instrumental in establishing self-restraint and control over the emotions.[6] The aim was to confine this arbitrary irrational portion of the human mind within surveyable bounds. Every educator of the eighteenth century saw it as his primary duty to gain mastery over a child's emotions and to shape them in a positive way. They might be applied at work or in society—in moderation. But the boundary at which they degenerate into erratic, unpredictable actions is a treacherous one: for order, virtue, and a properly regulated lifestyle, as the contemporaries saw it, were most immediately reflected in measured, composed bodily expression.

Gestures or Gebehrden

Like physiognomy, gestures or bearings *(Gebehrden)* also had potential as identifiers of virtuous conduct. For Johann Georg Sulzer, one's carriage was an expression of "what is happening inside the soul" (1771–74, 1:427 f.). A gesture was, according to Sulzer, a corporeal reflex linked to internal events. It was an indicator of emotive activity and could express this inner activity more directly and completely than words. "No words can define or express in

6. The inherent role of the emotions in the general feeling of happiness in life, as Plato claimed, was put into the background, and "praise for the passions" reemerged only in connection with the concept of genius (see Lanz 1971, 96). In the modern period the concept of affect *(Affekte)* combines "all nonrational—and therefore often understood as passive—mental phenomena" (p. 94).

so lively, let alone so rapid, a manner as the gestures, either passion or chagrin, contempt or love" (p. 428). More so than the countenance, gestures, or the way one carried oneself, were mainly ascribed a communicative character. Sulzer therefore concluded that nothing "can work upon the temper" as well as gesture. The pedagogue Johann Stuve identified in the fine arts an advantage for the educative use of bodily demeanor: "We know that verbal description can give us only a very dull, bland, and fragmented image of certain feelings and emotional states, which, in an instant, the whole outward appearance of a person directly displays with the greatest variety and liveliness and which spill over—so to speak—into us" (1788, 426). Bodily posture is one of the most important representational forms used by artists to express something and arouse feelings in a viewer of their work. But, "as definitely as each feeling . . . expresses itself by its own gestures, as indefinite and inadequate is every language, should one wish to establish rules for this portion of art" (Sulzer 1771–74, 1:428). Sulzer lamented this lacuna, and, because a set of rules for the repertoire of gestures still remained to be defined and no illustrated compendium of such poses existed, he suggested starting a collection of gestures.[7]

Gestures are an expression of a corresponding spontaneous inner emotion. They are also a communicative signal that emphasizes its symbolic character, which, however, does not necessarily have to agree with the instantaneous emotive event (see Kirchhoff 1974, 30 f.). Such significant gestures employ different parts of the body, whereby the hand is a particularly able symbol carrier and best fulfills a gesture's narrational function.

The discussion of physiognomy, deportment, and aesthetics in art, as sought by Sulzer and Chodowiecki at the Berlin Academy of Art, is also of relevance to acting, with a catalog of gestural signs serving as a useful reference. At the beginning of the eighteenth century Alexander Gottlieb Baumgarten proposed a theory of the passions *(Leidenschaftslehre),* and Gotthold Ephraim Lessing laid down firm rules for the eloquent role of the body. Eventually in 1785 Johann Jakob Engel undertook to recapitulate these thoughts systematically. In his book *Ideen zu einer Mimik* (1785–86) he classified the body language of actors in an effort to make expression on the stage more effective and authentic. To the question of whether the art of acting can be learned like notes off a music score or whether it must be performed instinctively, he responded with a compendium of different movements. Bodily gestures or mime had to be studied, and constant repetition was the best guarantee for fluency in the language of the body. Poses that exposed the operations of the soul fulfilled a dual purpose.

7. On gesture and its importance in cultural history, see Bremmer and Roodenburg 1992.

They served as representations of a *Bürger*'s inner emotions and therefore had to be carefully rehearsed by the actor. Moreover, they were worthy models for an audience to emulate and thus served as support for a spectator's memory (Engel 1785–86, 2:69). Quoting Sulzer, Engel asked, "Why should a collection of telling gestures be less possible and less useful than a collection of shell, plant, and insect drawings?" (1:70; Sulzer 1771–74, 1:428 f.) Just as naturalists profit from the latter, so also actor and spectator would draw equal benefit from a collection of gestures. "As for compartments in which to store and mount the treasures of your own observations, do not trouble to build them yourself. I am your man" (Engel 1785–86, 1:87). As an established index, gestural language entered the canon of virtues to be mentally and physically assimilated.[8]

The insight that modern forms of perception and human interaction have their historical roots in the behavioral norms of a bourgeois citizen, which developed with their associated physical and mental demands particularly in the eighteenth century,[9] is exemplified by the following representation of a *Bürger,* his "lines of sight" *(Blickrichtungen)* and his needs. At the same time it will be shown how little the education tools designed by these instructors actually fulfilled their original purposes.

The Depicted Person

Visual representations of sensory phenomena attained almost canonical status as a means of translating the standards of virtue into actuality (see Engel 1785–86, 1:57). The pictorial gesture, the typified human attitude in a copperplate engraving, painting, or drawing, linked theoretical moral concepts with their practical bodily execution. Chodowiecki was one artist whose work set out to combine aesthetics with moral notions. The influence of sense perceptions upon the understanding and reason was an intensely discussed issue: What were correct actions? How is correct behavior conveyed?

Perceiving a copperplate engraving is a sensory experience. Repeating such an experience is the basis for the cognitive faculty, and can be the way in which

---

8. A more comprehensive treatment of Engel's work on mime is provided by Bachmann-Medick (1989). In the context of popular philosophy of the late eighteenth century, she investigates the question "how [moral] action can be 'ordered' not just by rational activity, but by the aesthetic activity itself" (p. 7).

9. In this connection Herrmann speaks of a "codification of the bourgeois conscience" (1982, 153–62) and analyzes a *Bürger*'s moral and political education on the basis of C. F. Bahrdt's *Handbuch der Moral für den Bürgerstand* from 1789.

reason is implemented through the union of moral sensory experience and cognition. This is the reason that children's educators repeatedly asked for "good" copperplates. For, once imprinted on the brain, an image—the basis of all additional perceptions—is not so easily erased from a child's mind. Since children are just starting to perceive things and gathering their first experiences and, according to the sensationalist philosophy, the intellect and reason are established only at a later stage in child development, the first objects of perception are very important. Only through carefully designed representations of the human body could the external form serve as a model worthy of emulation. Or as Christian Cay Lorenz Hirschfeld formulated it in his speech before the Literary Society in Kiel on 11 October 1774 concerning the moral influence of the graphic arts, the visual arts "speak through the form, the features, the stance, the motion of the human body, which they present much more expressively than the orator's or poet's art. All that exists in the human character, the beautiful or the hideous, the sublime or the base, all this do these arts have the power to make apparent by means of facial expressions, poses, and gestures" (1775, 26 f.). He closed with the statement that images affected "the scholar and the ignoramus, the prince and the commoner; they are recognizable to all nations and speak a single tongue understandable to wild savage and lucid European alike" (p. 28). Realizing these required a sensible appreciation of a work of art, and, in accordance with a viewer's faculty of sensory perception, visual depiction could be a correspondingly powerful mediator between the outside world and the inner soul.[10]

Prints played a special part in this process. Originally serving principally as a means of reproducing paintings and for this reason duly placed near the bottom of the artistic hierarchy, this technique simplified a wide circulation much more than, for instance, the painting medium. As an illustration for printed text, it may at the same time create a link between word and image.[11] Prints experienced an extraordinary boom, which determined the publicity and popularity of the virtues they depicted. The "monopolized picture" (Hausenstein 1958, 9), that is, the original work or hand-painted miniature, represented the way of life and world of ideas of the nobility. A printed illustration, on the

10. The extent to which this was connected with a revision of the established formal language in the arts is described by Werner Busch (1977, 1993). The preexisting, strict, rational order to which the creative means had been subjected was upset and with it the prescribed manner of reading the pictorial works (Busch 1993, 20).

11. The classics in the history of book illustration include Hausenstein 1958, Lanckorónska and Oehler 1932–34, and Rümann 1931; see also the various contributions in *Die Buchillustration im 18. Jahrhundert* (1980), and Meier 1994, which interprets print illustration in the eighteenth century in connection with historical paintings.

other hand, could render common public themes or reproductions of already famous masterpieces and offer them for sale on a quite different scale of distribution. Pictures illustrating a written work, which were almost exclusively produced as copperplate engravings or etchings, performed a decorative function but were supposed to meet various other criteria as well. They could support the text as a visual guide during reading, or underscore the reader's impressions and allow the eye to participate in the process of comprehension. As reliefs, woodcuts, which were commonly used in the seventeenth century and occasionally also in the eighteenth century, could be printed together with the type. As intaglio processes, copperplate engravings and etchings required separate impressions (see Enklaar 1980), which led to the development of an independent branch of illustration printing (see Meier 1994, 9–15). In most books from the eighteenth century the engravings were either bound with the text, printed on loose sheets with the versos left blank, or collected at the end of the work. Each engraved copperplate could be used only for a limited number of impressions before becoming too worn and having to be refreshed or engraved anew. Such work was done not by the artist himself but by an engraver. Illustrations for books in high demand thus regularly had to be reworked many times. Although copperplate engravers did not like to regard their art as a craft, whence it had originated, their work was nevertheless performed in a workshop system (see Hoffmann 1934, 12). Daniel N. Chodowiecki, for example, employed a number of draftsmen and engravers for his work, and larger commissions such as the Picture Academy or the *Elementarwerk* were completed in stages. The drawing, plate chasing, printing, and, where applicable, reworking of the engravings were divided among the various experts.

The literary arts preferred smaller formats (octavo or duodecimo for ease of use and compactness). Larger folio formats were chosen for art books, atlases, and even primers like the Picture Academy. The function dictated the specific format: a novel enjoyed in a private setting presupposed portability and convenience, whereas a work intended for instruction was designed on a larger scale so as to enable many people to consult it together.

Illustrated books were one way of affecting people individually in an instructive way, and they quickly evolved into a teaching tool. In the course of this development, book illustrations split into two directions: "Whereas the idealizing direction strives by simplification to raise representation to the sublime or by typification to distill it to the mythical, *bürgerlicher* and petty-bourgeois realism with its painfully meticulous detail takes up Chodowiecki's artistic approach" (Lanckorónska and Oehler 1932–34, 3:6).

Daniel N. Chodowiecki succeeded first as an enamel painter and at the age of thirty-eight was accepted into the Berlin Academy of Art. He illustrated numerous works by Goethe, Lessing, Lavater, and Basedow and was one of the most sought-after portraitists of his day. One of the first artists in the eighteenth century to turn his attention to topics of the middle class, Chodowiecki naturally chose portraits as a favored subject—the classic symbol of the master and his intimate circle. In documenting the *Bürger* and his world, such works helped define a *Bürger*'s pictorial identity as well. Even the artist's self-portraits reflect the bourgeois values of duty, morality, and an unflagging work ethic.

Chodowiecki's statements reveal the hope that art would be comprehensible to everyone and that images should bear significance in a viewer's own life. Conversely, this newfound public offered the artist a certain independence from powerful clients like members of the government and the church. Illustrations should be designed in a way "that the heart itself is stricken by its moral message," which until then had only "been conceded to the theater as a public institution of citizens" (Beise 1992, 240). Thus, in a pictorial image, emotion acquired the public function of educator. It was no longer merely a potential reaction but should be roused as part of a plan to "stimulate" a person's imagination, "to warm his heart, and finally to stir his mind to reflection" (Hirschfeld 1775, 27).

The effect evoked in the viewer and that represented in the engraving are expressed in the posture and gesture of the *Bürger*. The importance Chodowiecki assigned to the language of the body is revealed in his suggestion to introduce a course at the Berlin Academy of Art to study the various expressions of the moods and—similar to Sulzer's call—to define a corporeal vocabulary for graphic expression. In his theatrical or milieu engravings he gives the exalted character an exaggeratedly animated demeanor or the arguer a gesticulative pose, yet the simple citizen is calmer and more staid, in conformance with accepted rules of conduct. The pedagogical effectiveness of Chodowiecki's plates is described by Johann Wolfgang von Goethe: "our trusty Chodowiecki has drawn many a monstrous, depraved, barbarous, and repellent scene admirably in such tiny monthly copperplates; yet, what did he do? He juxtaposed the hateful with the lovable" (1895, 294 f.). This illustrator's work is based on clear, black-and-white comparisons, whose simplistic moral associations (virtue is always attractive, vice is always repulsive) did not always win him praise.[12] Educators like Stoy, on the other hand, were happy to adopt this

12. Georg Christoph Lichtenberg, who had commissioned Chodowiecki for a series of contrasting themes, made the critical remark "What luck for virtue, would that vice were always so stamped! But, alas, it is not" (Chodowiecki and Lichtenberg 1977, 8). Lichtenberg missed the

simplistic manner of portraying morality: rudimentary education was allowed to use simplifying illustrations. Too much differentiation would be too complicated for children; the simple forms were most easily imprinted on the mind to facilitate understanding and emulation.

Gestures in the Academy

In the opinion of contemporary philosophers and practicing educators, imitation was the basic way in which social mores were relayed. Imitation, as it was understood in empirical psychology of the late eighteenth century, was a nature-given form of behavior. "The impulse to imitate is formed early on, and its origin seems to lie in a mechanism that assists in the association of ideas; whenever we think of something instructively and observe the movements of another with full attention, we cannot but do likewise to our utmost abilities" (Tiedemann 1787, 322). The copying impulse in children was, accordingly, geared not only to the behavior and employments of the adults around them but also to the illustrations of moral and civil actions by characters they saw in pictures. Pedagogues of this period were unanimous in requiring that a child be shown only the right things and occupations, to assure that the impressions formed in a pupil's memory were guided along the proper paths from the very start.

In his text to the Academy, Stoy also worked with contrasting pairs worthy of emulation: good versus bad, beautiful versus ugly. "*Education of children* consists in habituating them to discerning the useful from the harmful; to doing one thing, and leaving the other be" (1780–84, 371, original emphasis). That is how Stoy began the chapter about the thought and care underpinning the effective instruction of children. This part of the Academy, its fourth field, was reserved for the *Elementarwerk;* and this statement, which Basedow had worked into a longer essay about the family and extended relations, Stoy used as a programmatic introduction to his chapter. Discerning the useful from the harmful accordingly required a clear segregation of the world by category. Both are very clearly defined in the Picture Academy. Its readers, parents and educators, easily found their way through the polarities of good and evil, right and wrong, light and dark.

Accordingly, children are basically innocent, happy, and good (see p. 24). The "higher faculties" like intellect and will, which lead to "thoughts, con-

---

requisite differentiation precisely in the illustrations of polite behavior, which rely heavily on such visualized opposing pairs. Compare William Hogarth's illustration "Analysis of Beauty," which appeared in 1753 and served as a model for Chodowiecki's engravings (see Kemp 1975, 120 ff.).

cepts, judgments, and rational associations" (p. 33), are, according to Stoy, contrasted to the lower faculties of "sensible perception" *(sinnliche Erkenntniß):* "these include the senses, the imagination, and the memory." These also include the "desires of the senses" or passions (*sinnliche Begehrungs-Vermögen,* pp. 33 f.), that is, the good emotions (hope, or even joy) and the bad ones like jealousy, greed, and revulsion. It was a matter of degree, however; although sense perceptions were formed from the more basic faculties of the mind, it was only through applying the reason together with the senses that true insight might be gained. Good conduct always required equanimity, good humor, and confidence; bad conduct always involved restiveness, anxiety, peril, and retribution (pp. 58 f.). He who lived happily, lived tranquilly as well (p. 155).

Besides anger, fright, and sadness, Stoy regarded "violent *motions of the body*" (p. 569, original emphasis) as the source of bodily affliction. "Thus an excess of these and similar *agitations* robs the mind, the senses, the mores, and a person's whole lifetime of energy, truth, purity, and permanence; whereas *composure* and *moderation* help gather, increase, and safeguard these riches of the mood" (p. 665, original emphasis). Some good might be found in emotions like sadness and joy, but only when "their measure and purpose is prescribed by reason" (p. 663). And Stoy concluded with an important message from personal experience, which he expressed in verse (p. 565).

> When inside you passions rage,
> Learn to command their vanity, their power—
> Offer up all with joy to your God,
> Passions, fleeting pleasures—your whole life long!

Useful work meant busying oneself steadily in quiet seclusion like a silkworm: "But his cocoon gets stouter by the day, and from that it can indeed be taken, that he is not idle in his hideaway" (p. 292). Harmful things, on the other hand, expressed themselves in erratic motions: a man who cursed too much could expect an ignominious death with "his look and actions wild" (p. 320).

The above polarities have metaphorical and symbolic motion analogies as well: good remains staid and calm, evil is loose and unruly. Thus Stoy admonished one girl: "This foolish maid is leaping too high" (p. 859). Coupling contrasting elements also had something to do with the aesthetic sense, according to Magister Philoteknos in Christian Felix Weiße's *Kinderfreund:* "In my view, good taste constitutes the ability to sense what is good and what bad, what is beautiful, mediocre, and hideous, and to discriminate surely between them" (1780–82, pt. 21, 274). Since hideousness was always correlated with indecency, disobedience, and a "dissolute way of life" *(liderliche Lebensweise),* a child's education in matters of taste and judgment was highly important, and visual depictions were an inexhaustible aid in this. A child's handling of aesthetic things was here inextricably linked with moral aims, and a piece of artwork was looked upon as a visualized lesson in deportment.

The Picture Academy's copperplates also used this device of opposites. They are usually expressed in gestures and offer a means of orientation among the 468 individual engravings. A person who "through a bodily ailment or vehement passions has completely lost his mind" (Stoy 1780–84, 61) is represented with a furious demeanor and "flowing hair." That is why the passage about the oracle of Delphi in copperplate 17 (plate 12) describes the divining priestess as follows: "One saw how her hair stood on end. She had a wild look in her eye, she was foaming at the mouth, her whole body began to tremble violently, and she was subject to all that which befalls a mad person in a raving frenzy" (p. 296). The priestess exhibits all the external traits of a heathen: arms flailing, mouth gaping with horror, hair loose, breasts bared, apparently tumbling from her seat. There is no mistaking who is on the good side and who on the bad. The Furies are similarly wild, with a shock of snakes writhing upon their scalps. Their movements are expansive and unbalanced (plate 13). The three Graces, by contrast, stand erect in classical poses, with placid expressions. Basically, the visual imagery into which good and evil are translated to illustrate the Academy's values always places excessive and appropriate behavior side by side.

The picture of a sapling that had not been bound to a support in time and grew up crooked is a plain metaphor for a badly bred person. The farmer or orchard grower is not able to pull the straggling tree straight anymore (plate 14). He had managed to straighten only the younger trees at the back. This juxtaposing of motion and motionlessness also applies to the wise man and the fool (copperplate 47, field 9); and Hercules likewise must choose between beckoning vice and upright virtue (copperplate 3, field 8).

Besides these obvious contrasts, the Academy also uses more subtle forms: In a scene apparently depicting a business transaction, a smart person is set opposite a married couple who, according to the motion versus motionlessness criterion, are acting inappropriately (copperplate 3, field 9). And the emblematic depiction of a smoothly flowing river and a torrential current—paradise versus wilderness—is based on this same contrasting scheme. The contrast between these persons or conditions is revealed in the open arms, wide gait, splayed feet, exaggerated facial expressions, and bent postures for the bad characters. The good characters, by contrast, take a sedate and relaxed stance, feet drawn close together, arms usually against the body, a balanced pose with head slightly uplifted.

The gestural repertoire used in these illustrations resembles Chodowiecki's in a series of copperplates for Lichtenberg's Göttingen pocket calendar for 1779 and 1780, entitled "Natural and Affected Attitudes of Life" *(Natürliche und affectirte Handlungen des Lebens).* This series of engravings compared the middle-class (natural) attitudes of a *Bürger* with aristocratic (affected) bearings and for this purpose principally used body language. The opposing pair labeled "The lesson" will serve for comparison with the Picture Academy's plates: they portray the typical deportment of *Bürger* (plate 15) and nobleman (plate 16).[13] In these examples, a "natural attitude" repeats the same good gestural qualities: a passive, erect posture, with arms folded or held against the upper body to form a uniform whole. Where visible, the faces emanate a quiet, even temper. An "affected attitude," on the other hand, takes possession of the space: flamboyant arm movements, parted legs, and pallid bored expressions; the striking style of dress and hair accentuates the haughty air of the figures.

Comparisons between the middle and upper classes and a *Bürger*'s identification with disciplined bodily expression have been the subject of many studies.[14] Taking all the qualities of the good *Bürger,* as opposed to the bad or stupid one, in search of a prototype of each in the Academy, plates 17–20 provide model bodily stances. The classic *Bürger* stands with a straight but not stiff back, his weight consistently shifted onto one leg, and his arms resting against his body; he sits attentively and upright. One hand is either raised in admonition, pensively folded with the other, or tucked into his breast pocket. This "virtue of quiescence" *(Tugend der Lässigkeit)* reflects the self-confidence of an eighteenth-century *Bürger,* his frame relaxed, yet as though strung within a muscular shell (zur Lippe 1974, 2:238). Everything about him looks poised and alert yet, despite the attested "mean motion" *(mittlere Bewegung),* without dynamics. Every copperplate demonstrates how, with head held high, one maintains a commanding view over the various situations in life. A *Bürger* has nothing in common with simple vagrants or with the sick in a hospital ward (cf. Stoy 1780–84, copperplate 43, field 6). Standing clearly apart from the lower echelons, he advertises by his civil composure his authoritative proprietorship.[15]

---

13. Among the many descriptions of this series of engravings, cf., for instance, Barta 1987, Gallwitz and Stuffmann 1978, and Kemp 1975.

14. See Barta 1987, Barta-Fliedl and Geissmar 1992, Böhme and Böhme 1983, Duden 1977, Elias 1976, and Kemp 1975.

15. Edith Hoffmann confirms this model in the painting tradition of the eighteenth century. She notes a lack of "action" among the figures and a "new, quite undecorative stiffness" (1934, 24).

The Graceful Bürger

According to Johann Caspar Lavater a consummate person combined calm dignity with commanding power and humble wisdom. During the eighteenth century these characteristics were united in the typified ideal, the *Bürger* (see Louis 1992, 127). The Picture Academy's representation of this type comes very close to Lavater's ideal, which he compared with the charm and balanced proportions of Greek statues. Like his contemporaries, Lavater considered Greek antiquity the definitive epoch, which, as Winckelmann saw it, was characterized by "serene simplicity and noble stature" *(stille Einfalt und edle Größe).* The beauty ideal of classical Greece had already been embraced in the seventeenth century by Charles Le Brun, who used it as a basis for a kind of repertoire of gestures and theory of proportions for the human body (Louis 1992, 121). Appreciation of ancient Greece was revived again in the mid–eighteenth century, and the concepts of taste and proportion, of such importance in education, were defined according to this ideal.

In this respect Winckelmann's description of one statue reads like a synopsis of the ideal *Bürger:* "The more sedate the condition of the body, the better is it able to reveal the true character of the soul. . . . The soul is more recognizable and expressive in violent passions, yet it is great and noble indeed in the pose of harmony, in the pose of rest" (1995, 31). The "pose of rest" does not necessarily have to lead to a rigid, straight line or stiff, erect posture. William Hogarth's plate "Analysis of Beauty" from 1753 describes "the line of beauty" as a medial line, neither crooked nor quite straight, neither a caricature nor wholly without expression. In a famous depiction of a fictitious sculpture court in which an ideal collection of typical statues is assembled, a dance master is standing by the figure of Adonis. Next to the graceful Greek, the dance master looks ridiculously stiff; and by extending his arm—as if to correct the statue's pose—the caricature is complete. "Mean motion" (*mittlere Bewegung,* Gallwitz and Stuffmann 1978, 34) or moderate graceful movement may be added as another ideal prototype to the canon of poses. "Mean proportion" *(mittlere Maß)* is one of the guidelines of eighteenth-century art. Peter Gerlach agrees that "the history of proportion is not merely a reflection of the history of style . . . , but in these reflections basic anthropological patterns of a far-reaching kind have been defined" (1984, 74). Thus the contemporary anthropological discussion and artistic tradition of representation meet among the educational notions of the late eighteenth century. Both Hogarth and Daniel N. Chodowiecki made an appeal for a gestural language different from the aristocratic tradition of ecstatic sensitivity or exquisite rapture; what characterized a *Bürger* for these two artists, just as for the Academy, was the "frugality of gesture" (Kemp 1975, 122).

The figure of the *Bürger* and its canon of forms[16] was drawn from various lines of tradition. To the ideals of Greek antiquity and the theory of proportion with its mean motion, and to the well-formed body and symmetric face is added the depiction of the prominent forehead to represent the ideal of intelligence *(Geistigkeit)* in the cultivated mind.[17]

The head raised slightly with chin protruding indicates a claim of authority. The convention of the "raised head" (*erhobenen Hauptes,* Warnke 1992, 190) originates from the profane exhibition of power, pride, and merit. This head position, indicating the ability to observe and survey with keen penetration, is included in the Academy's repertoire.

Another gesture of the display of power is the hand resting on the hip with elbow bent. Joaneath Spicer traces this gesture back to the fifteenth century in paintings. From the beginning it typically served the "male military figure registering self-possession and control, either the assertion of success or defiance" (1992, 86). Usually leaders and people in command are depicted in this way, standing with one leg carrying their body weight, the other leg relaxed, which signals complete composure but also an element of personal display. Another gestural borrowing for the figure of the *Bürger* is the hand brought up to the head or the head held in hand, as in the classical pose of the thinker. Correlated with a moment of contemplation, this pose signifies the engendering of knowledge and art: it signifies mental activity. Warnke attributes the new way of representing hands to the changed role of artists in the sixteenth century. The peculiarity of an artist's hand was the source of his uniqueness. The hand of an artist "should be made to acquire through practice the skill to receive and translate the thoughts of the mind, to short-circuit head and hand, so to speak" (1987, 57). Thus the *docta manus,* the learned hand, established itself, no longer merely copying the world of objects but shaping it as well. The hand of the *Bürger* too was assigned an intellectual function replacing that of a pure

---

16. Using the example of Hogarth, Busch (1977) describes what place the inherited canon of forms takes in the "*Bürger*'s principle of art" and its shift in meaning in the eighteenth century.

17. The brow—a fundamental constant in facial proportions—played an essential part in the anthropological debates held in the eighteenth century about the purported superiority of Europeans. Pieter Camper's studies of the "facial angle," which is formed at the intersection of the line extending from the outer ear opening to the nasal cavity area and the orthogonal line connecting the brow prominence with the outer extremity of the upper jaw, were the basis of his craniometric analyses. As Visser (1990, 330 f.) writes, they were not yet used as a manifestation of inferior ranking of non-European peoples; Samuel Thomas Soemmerring was the first to take them up as a basis for a hierarchical classification of the races. The "high forehead" was given special significance in the eighteenth century as a measure of intelligence (see also Louis 1992, 121).

craft. “Touching one’s own head . . . is a sign of intellectual strength. In this pose the figure becomes a ‘victor of thought’ [*Denk-Sieger*]” (Barta-Fliedl and Geissmar 1992, 70).

The Picture Academy shows how in the eighteenth century a variety of preexisting forms of expression were drawn together and melded into a single prototype. The result—to borrow Aby M. Warburg’s expression—could be called a “vulgar Latin of the language of gestures” (Gombrich 1970, 251). In the Picture Academy, pictures are an instrument not of orientation but of identification of temporally specific values of self-restraint, moral diligence, and encyclopedic knowledge. But the new content attached to gestures already laden with different meanings produces a rather comical effect in many instances, instead of the intended clarity, due to the diminution of the pictures and their placement among entirely differently directed motifs. The print medium and its possibilities are at fault here, as is the attempt to attach an artificial didactic moral to a specific picture and its gestures.

If the surgeon’s assistant with his graceful manner is able to stand firm before the violence of Cain or Romulus, a man-eating crocodile, and a stalking wolf (see figure 2), the cultivated man of European stock in field 3 (plate 21) of copperplate 2 looks downright peculiar directly above the clumsy baroque scene of Adam rushing forward to embrace Eve in Paradise. The intended opposition of motion versus motionlessness is scarcely noticeable; in copperplate 31 (plate 22), the primates in field 5, together with all the unsettled figures in fields 2, 7, 8, and 9, seem much more interesting than the settled figures in fields 1, 6, and 9. In copperplate 30 the sheer number of images and themes completely overwhelms any sense of contrast, and the eye is distracted by other things. The classical repose of the Virtues in field 2 of copperplate 34 looks almost ludicrous set right above a ferocious rhinoceros.

These few examples suggest that the Academy’s way of representing the middle-class *Bürger* is insufficient, not least because the tiny illustrations in the individual fields can express the gestural formulas only schematically and the general context of each copperplate in which they are embedded frequently conveys contrary messages. A review from 1780 noted that the correlations between the motifs on a tableau were “forced, at times” (*Allgemeine Bibliothek* 1780, 8:385), and ended up “too small” (p. 388). Moreover, the combination of encyclopedic knowledge with representations of moral attitudes results in a conflicting pictorial effect with split purposes: whereas the pictorial transmission of encyclopedic knowledge depends on faithfulness to individual detail, mores and virtues rely on the reiteration of ever the same patterns.

Much significance is given to gestures in the Academy, and Stoy's confidence in the moral effectiveness of copperplate engravings was a guiding motive in its conception as well. Nevertheless, the simple opposition between active, sweeping bearings and static, clearly structured ones ignores their origins when placed in conjunction with the representational formula for a *Bürger.* Nor do they fulfill their dual function as plain ethical signs and formal organizational aids in the acquisition of knowledge. How moral instruction and the accumulation of knowledge can mutually exclude each other and how thematic overlaps between them can produce ambiguities of content is illustrated by the following picture comparison. The gesture of arms held high and the mime of astonishment signals not only the trance of the divining heathen (plate 12) but also the anguish of a good woman, likewise bare-breasted, who is startled from her prayers by the building collapsing about her (plate 23). Judging from the expressions of the two women, oracle and prayer are indistinguishable. Yet it is exactly this that a differentiated mime ought to have achieved: a distinction between the confused mental state of an oracle and the humble submission of a pious woman to her fate. The small format of the Academy falls short, and the gestures remain ambiguous.

Here a pedagogical theorist offered the whole gamut of gestures and motions, to which he associated specific bits of information. Yet his attempt at simplicity failed. The images employing the orientational device of motion versus motionlessness are unable to reflect the intended one-dimensionality in meaning, largely because of the extent to which printed art could be proliferated and duplicated. Particularly in the case of the Picture Academy, the immense number of collaged picture excerpts and individual illustrations—by a variety of draftsmen and engravers—prohibited the desired perceptual uniformity. The diversity of themes alone leaves room for inconsistencies. The engravings are unable to render the nuances in meaning contained in their explanatory texts.

In arranging his Picture Academy in tableaux Stoy had originally planned to create an order among the separate engravings so that they would illuminate one another. But closer examination reveals that this juxtaposition was executed on a more superficial level with a practical, mnemonic aim. The result is, as a consequence, forced—to borrow the reviewer's term. Actual use of the individual fields reveals how little the sense of their tableau is geared toward reading the pictorial messages of the separate engravings. The viewer is kept busy just identifying the relation between the images and weighing this combination against the information in the accompanying text. Added to this ambivalence between content and pictorial presentation (as was exemplified by copperplate 38, the miracles of Christ) is the unclear message of the pictures themselves, particularly as concerns gesture.

The Visual Person

In Stoy's design, the sense of sight is the fundamental basis of a child's perception. The following account from Adelung's weekly *Leipziger Wochenblatt* illustrates the types of vision that educators as well as naturalists and painters were aware of in the second half of the eighteenth century.

It reported the experiences of a youth who had been blind throughout childhood and first gained his sight as a result of a cataract operation. Because sight "requires much experience and many decisions of the mind, when we wish to form a proper idea about the things that we see," it was particularly difficult for a fourteen-year-old boy to pick up what was normally experienced and learned during infancy. "When this boy's glassy gaze was taken away and for the first time he began to see, he could not conceive that the things that he saw were so far away from his eyes; for he imagined all objects were touching his eyes, as what he could feel was touching his skin, and he found nothing so pleasant as objects that were smooth and regular." The boy slowly had to learn how to look at and understand paintings and how it is possible to represent something three-dimensional on a flat surface. "At the outset he could survey very little at a time and judged what he saw to be very large. But when he saw bigger things, he imagined the aforeseen things smaller." The story ends with the final exercise in the training of his sense of sight: "One year after he had obtained his eyesight, he was brought to the top of Epson Hill. There he saw a wide view, took inordinate delight in it, and called it a new way of seeing" (*Leipziger Wochenblatt* 1775, 7:113–18).[18] This account, culminating in the boy's being led to a hilltop to enjoy the panorama, accentuates the comparing and surveying look.

A few years later in his essay on the schooling of the body, Peter Villaume lamented the neglect in education of training the senses. When someone looks at an object for the first time and is not practiced in discerning things at a single glance, then in his or her mind's eye all the parts of the object "swim together into a single confused perception" (1787, 474 f.). For this reason children should be taught the right way to see; that is, they should be able to

18. Stories about the blind regaining their sight are a legendary topos similar to that of the artist (see Kris and Kurz 1980). In the eighteenth century cataracts were regarded as unusual, and successful treatment of them was not necessarily the rule (Stoy placed it among the wonders). This episode also addresses the importance of the senses of touch and sight (Morgan 1977; cf. Immerwahr 1978 particularly on Diderot and Herder).

determine distances and estimate or assess things against each other (see also Stuve 1788, 233 ff.). Sight ought to be made more definite and reliable, because all the insights that children had gained until that time were incomplete and full of misconceptions (Villaume 1787, 474). And Villaume continued: "It is not enough that hands and fingers be sound, strong, and sturdy; they must also possess a certain dexterity and skill. Not enough that my eye be bright and see acutely . . . : it must be able to survey much at a single swift glance, so as to be able to decide rapidly and act accordingly" (p. 285). It is the comparing, discriminating eye that composes an "overview" *(Über-Blick).* These forms of vision place observational sight at a more differentiated level than simple alertness. Looking should be done not merely with intelligence, but with trained insight as well. The tools for understanding objects should be made available, and with their aid a specific form of perception could be trained. The boy's successful recovery with its dramatic climax and conclusion on the hilltop constitutes a legitimation for such schooling of the eye.

The Academy's picture tableau answers Villaume's call to train the eye. The copperplate tableau requires not only a comprehensive survey of the essentials but also a comparative, ordering eye.[19] Such training was not yet feasible in Comenius's *Orbis pictus,* with its individual illustrations—nor was it necessary. All that is needed there is the attention required to understand the subject and confidence that the eye can convey something sufficiently real. After Descartes drew attention to the senses, studies on optical illusions and sensationalism warned that the senses could deceive but that one could override them by using them in combination.[20] Here for the first time educators systematically demanded instruments to impress and train the senses, and devised such instruments as well. On the basis of one sheet of illustrations it may be shown that, in the eighteenth century, in order to identify an image, the "eye" *(Blick)* was challenged to consult a dictionary—in other words, it should itself be able to perform an encyclopedic categorization (see te Heesen 1997).

---

19. Besides the points raised in the following description of the picture tableau, other elements guiding the eye also exist: Stoy fuses text to image by means of a referencing system of ciphers and lowercase letters that are explained in the text. The Academy's illustrations thus fall under the category of the factual figure in technical and medical literature. Just as the pointing line, introduced into anatomical incunabula during the fifteenth century, is supposed to facilitate clear description and to point the reader to the essence of an illustration, the cipher here is the "lineless pointing line" *(Hinweislinie ohne Linie)* or the reduced form of directing the eye (see Herrlinger 1967, 62 f.). Another such element is taken from the manner of illustration in Diderot's *Encyclopédie,* in which the copperplates pull each subject out of its context to represent it against a purged background (see Barthes 1964, 11).

20. Descartes also thought the senses were deceptive, but he saw their corrigibility rooted in the human soul; only the soul could lead one to insight (see Hagner 1990, 20).

The Multifield Picture

A sheet of illustrations, or the arrangement of many images on a single surface, already was a familiar medium for graphic depictions of Christian salvation.[21] In tracing back this compositional grouping one must make a distinction between two fundamentally different types. In one form the multifield grouping is arranged around a center that fills the largest space on the sheet (cf. the Picture Academy's schematic subdivisions). In the other form all the fields are the same size and are arranged sequentially. The first form frequently focused on a person (in homage). Lacking a central field of focus, the second fulfilled a more informational, narrational function.

Both forms are distinguished less by their sources or places of origin than by their functions. While the former is better suited as a religious medium for objects of worship,[22] the latter invites the viewer to read and combine the various images.

The "accounting, recounting" *(erzählende, aufzählende)* form of multifield pictures has been used for educational instruction since the infancy of printed art. These sheets of "ordinary prints" were generally designed for memorization. As a rule, explicative lines of text were added to the narrational sequences of pictures, or passages of text running along the edge of the sheet finished the story. Emblematic catechisms like that of Johann Adam Behr from 1718 adopted the centralized form for their illustrations. At the center of the illustration for the sixth commandment, Jesus is depicted delivering the Sermon on the Mount. This was supposed to illustrate that "purity of the heart" is the precondition for obeying the commandment; the peripherally depicted scenes relate to this central image (see Schug 1988, 146). The growth in number of such graphic illustrations in the seventeenth and eighteenth centuries is explained by the rising importance of children's education, a new emphasis on the visual sense, and the associated tendency to present knowledge in a transmittable form.

A further change emerged in popular graphic design as well. The multifield image of the eighteenth century emphasizes the ability to compare the individual images and their motifs, not just their narrational function and their registration "at a single glance" *(auf einen Blick)* but also the "comparative look" *(vergleichender Blick),* which darts back and forth between the individual components before a judgment is formed. The vignette Chodowiecki

21. Kunzle's history of the comic strip (1973) provides an elucidating comparison to the multifield picture.

22. A good representative of the centralized form is the biographical icon of John the Baptist from the thirteenth century (see, e.g., Belting 1991, 285).

designed for the title page of Lavater's *Physiognomische Fragmente* of 1775 contains a tableau inviting the eye to draw comparisons (plate 24). In order to sharpen the sense of judgment of the reclining youth before her, Nature is showing him a series of heads and physiognomies. Taken together they are supposed to represent a uniform whole; a practiced eye is required, however, to assemble the separate fragments. Here the ordering and comparing gaze appears in a reciprocal relationship. Comparative study by means of pictures makes it possible to juxtapose very diverse objects, and this method became important primarily in the eighteenth century as a way to engender insight, drawing its inspiration from the study of nature. The topos of comparability gained significance in natural history as it was applied toward the goal of an all-encompassing classification of the natural realm.

Stoy too aimed to school this "comparative seeing" *(vergleichenes Sehen)* with his Picture Academy. In the copperplate volume each sheet is a centralized multifield illustration, although the focus here is on Holy Scripture. Hence, this person-oriented representational device was chosen to illustrate a text: it is not the aura of a person but the powerful influence of the Bible from which potential points of reference should be sought to link the peripheral fields. Unlike Behr's catechism of 1718, however, this pictorial cross-referencing system, whose original application was to Christian and biblical contexts, was used to link worldly bits of knowledge whose intrinsic informational content is quite unassociated with the biblical leitmotiv.

The centralized display, which possesses a considerably greater power of suggestion than a simple series of pictures, seems to disintegrate, however. In Stoy's day it was no longer a matter of course to refer back to a single focal point, let alone to a biblical one. In the general discussion about natural history, for instance, monkeys take up more room than could ever be warranted by the central field of the Academy's copperplate 31, which depicts the biblical episode of King Nebuchadnezzar. This canonical form of grouped pictures, which harks back to the beginnings of Christianity, is filled here with foreign content that it cannot comfortably accommodate. The established form of serial grouping of pictures around a center dictates a special regard for the central material. The Academy's tableaux must be read in this sense, with biblical events serving as the central point of reference. But the accompanying text assigns equal status to all the fields, and the focus serves merely as a linking element. At places it seems as though Stoy was aware of this problem (plate 22): copperplate 31 does not even have a proper center to its structural subdividing. Broken into three parts, the middle field is reduced to a narrational series of images that no longer can claim absolute thematic dominance. What first attracts the eye are the monkeys' antics on the left-hand side of the copperplate. With their sweeping movements against a light background, they stand out as the clearest segment on the page.

Picture arrangement and construction of the signifying content are crucial for the effectiveness of an image and a viewer's association with it. The motif itself is not the only vehicle of meaning; its presentational form is equally important. In planning the central focus as Holy Scripture and allotting it the largest central field, Stoy nevertheless allowed the presentation to undermine his scheme.

Graphic composition and the juxtaposition of various thematic traditions led to a heterogeneity in the pictorial effect that was certainly not Stoy's intent. A closer examination shows, moreover, that knowledge and morality cannot appropriately share the same context in the Academy's tableaux. The eye, intended to be schooled in alphabetical perception of the world, in orderly appraisal, fails from the opaque centralized form. Just as references to the divine sovereigns and rulers retreat to the forewords of scholarly works, so the Holy Writ must cede its place as the center around which all knowledge revolves, even in children's literature of the late Enlightenment. With the inability of a picture tableau to communicate directly between observer and the world of perceptions, Stoy's centralized multifield picture had lost its material and formal focus. Into a pictorial form that had hitherto met the needs of substantially uniform content is injected the building-block principle, juxtaposing heterogeneous, equal-ranking levels of reality. The much touted "comparative seeing" cannot be trained by means of the Picture Academy, according to the pedagogue's original conception, just as its repertoire of gestures does not provide the requisite clarity.

The ambivalence between the visual and textual messages conveyed by the Picture Academy, the intrinsic ambiguity in its depictions, and finally the cross-purposes between the illustrations and their presentation are major reasons for its flagging reception in later years. Good sales for the first two installments subsided, and many customers may not even have completed their sets. Once again Stoy's picture book was eclipsed by its more illustrious rivals published by Basedow and Bertuch.

Tableau B

Plate 12. The oracle of Delphi (Stoy 1780–84, copperplate 17, field 8)

Plate 13. The Graces and the Furies (Stoy 1780–84, copperplate 34, field 8)

Plate 14. Straight and crooked saplings (Stoy 1780–84, copperplate 21, field 7)

Plates 15 and 16. *Natürliche und affectirte Handlungen des Lebens:* "The Lesson" (1779) (Focke 1901)

Plate 17. A tailor and his customer (Stoy 1780–84, copperplate 17, field 6)

Plate 18. A family listens to a hermit (Stoy 1780–84, copperplate 28, field 9)

Plate 19. A man warns a boy (Stoy 1780–84, copperplate 37, field 9)

Plate 20. Encounter with a vagrant (Stoy 1780–84, copperplate 37, field 7)

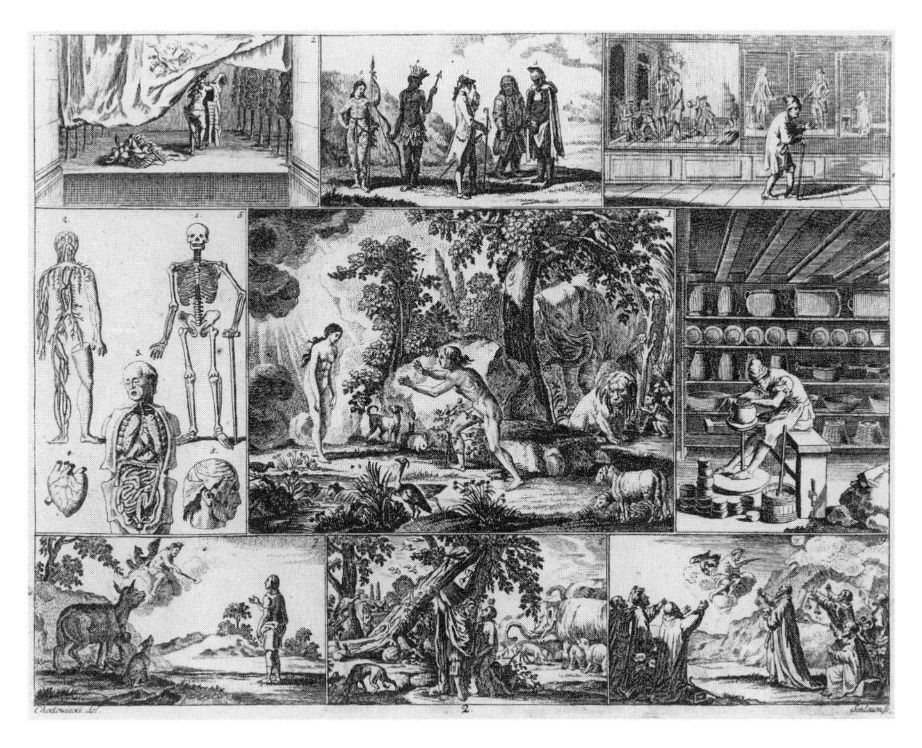

Plate 21. Copperplate 2 of the Picture Academy (Stoy 1780–84)

Plate 22. Copperplate 31 of the Picture Academy (Stoy 1780–84)

Plate 23. Praying woman in a collapsing building (Stoy 1780–84, copperplate 27, field 9)

Plate 24. Title-page vignette for *Physiognomische Fragmente* (Lavater 1775–78, vol. 1)

BOX

5

THE BOX AND COLLECTING

Johann Siegmund Stoy presented the Picture Academy for the Young either in picture-tableau form or as a boxed set of engravings. In this variant the tableaux are cut up into nine picture fields, and the resulting cards are sorted into compartments arranged in the same manner as the fields in a tableau (see figures 1 and 4). The pictures of a single category are grouped together, and the user has direct access to all the picture cards of a single thematic field without being compelled to determine their relation to the biblical world. For this reason the box complies better with Stoy's encyclopedic demands than an uncut picture tableau does. Card catalog–like structure is nothing extraordinary for this "classifying age." It served merchants, naturalists, and teachers alike as a means of organizing their knowledge. Against this backdrop, Stoy's Academy should be regarded less as a historical children's book than as one object among many used to further information and learning. Consequently, the box form lends new significance to what has thus far been presented as an ambitious illustrated schoolbook. It formed a part of the material culture of the eighteenth century.

The Memory and Its Places

In his preview to the Academy, Stoy used the recurrent metaphor of mental repositories. Thus he called the Picture Academy "a store of good food for the mind" (1780–84, directions, 1) with "materials for the instruction of the young" (foreword, 1), or a "storehouse of the finest materials for education" (p. 10). Thanks to his arrangement and compilation of the individual themes, youthful minds would be enriched with "immeasurable treasures" (p. 11), and this collected "stock" (p. 12) would serve as "exercise for the memory" (directions, 6), in which to "build particular receptacles" (p. 8). The Picture Academy was an *Universalrepertorium* (p. 8); and just as an "orderly housekeeper puts everything away in its proper place" (p. 10), one should proceed likewise with the "gems of learning" (p. 10), so as to enable easy retrieval of what has been set aside in the memory. Stoy mentioned that the copperplates should serve a child's imagination as "a well-furnished *repertorio*—all that which [a child] hears, reads, and sees at the same or a future time, shall be easily set down, as it were, alongside a main or adjunctive idea of this book, signaling itself either by the memory or by the quill" (1782, 4). Here Stoy uses age-old metaphors for mnemonics or the art of memory, which usually employ a physical space to symbolize the memory. The trick of those proficient in this art was to use memory slots and compartments for mental orientation and retention of a specific idea. So metaphors like "treasury," "theater," "archive," and "library" are common. But terms like "stockroom," "magazine," and "storehouse" as Stoy used them reflect a particular professional world. They point to the businessman and merchant.[1] Stoy's purpose was to summarize the "gems of knowledge" and to organize them within a "main plan" (1782, 2). Just as in constructing a memory space, devising a main plan means providing the knowledge or memorabilia with a taxonomy.

In his essay *De oratore* (book 2, § 86, 354) Cicero points out that rational organization is crucial for a good memory:

> Persons desiring to train this faculty must select localities and form mental images of the facts they wish to remember and store those images in the localities, with the result that the arrangement of the localities will preserve the order of the facts, and the images of the facts will designate the facts themselves, and we shall employ the localities and images respectively as a wax writing tablet and the letters written on it. (1988, 467)

The art of memory and recollection basically involved imagining a building or mental place with a fixed interior. Into these loci the *imagines agentes* are deposited. These images are emblematic and contain within their simple structure the content that the orator has previously given them for memorization. When delivering his speech, he can "pace through" this furnished memory chamber and systematically retrieve each image he has stored there with its associated content.

It is critical that the mnemonic classification make sense and be well ordered. In this way the memory becomes a knowledge repertory that the speaker can access at will. Each separate bit of information should have more than one

1. A survey of the various metaphors for the memory and their development into the twentieth century is provided by Hartmann (1989) and Assmann (in Assmann and Harth 1991, 13–35). Yates was the first to devote a study to this topic (1999; first published in 1966); the anthologies Assmann and Harth 1991 and Berns and Neuber 1993 describe the forms and functions of the memory and its social and cultural significance. The history of psychological and neurobiological concepts of the memory is summarized by Florey (1993).

link to others, so that a dense network of mutually supporting interconnections is formed. Because any one image or locality can carry more than one meaning, the memory chambers may be traversed in a variety of ways.

With the invention of movable type and printing of artwork in the fifteenth century, the role of mnemonics changed. In this regard Berns refers to a "revamping of mnemotechnics" (1993, 35) in which the Reformation, Protestant criticism of art, and above all developments in the history of media were instrumental in disseminating mnemonic pictures and writings. Mnemonic pictures or morally loaded images persisted into the eighteenth century and are rooted in old imagery traditions.

What Stoy ultimately fell back upon is described in the Thomist rules of mnemonics: "The first of these is that he should assume some convenient similitudes of the things which he wishes to remember; these should not be too familiar. . . . Secondly, it is necessary that a man should place in a considered order those (things) which he wishes to remember, so that from one remembered (point) progress can easily be made to the next. . . . Thirdly, it is necessary that a man should dwell with solicitude on, and cleave with affection to, the things which he wishes to remember. . . . Fourthly, it is necessary that we should medi[t]ate frequently on what we wish to remember" (Yates 1999, 74–75). Good organization of the things to be remembered, persistent occupation with them, constant repetition, and unusual signs or pictures constitute this didactic program, which Stoy propagated on the pages of his Academy and which his copperplates put into material form. This memorizational approach at least has hardly changed.

Places in the Art of Mnemonics

In drawing on the art of memory, one of the primary goals of the Picture Academy was to find a form suitable for the instruction of children. For a better grasp of the *ars memoriae* and Stoy's didactic intentions, it is necessary to take a closer look at the concept of collecting and its physical expression since the Renaissance, in the form of the cabinet of arts and curiosities. There is a connection between the mnemonic art and collecting on the one hand, and between the Picture Academy and collections on the other. The history of collections may illuminate the Academy's aim and show how it engenders insight and knowledge. The connecting elements are the loci, the places, which in the process of accumulating knowledge are materialized in the chambers and later in the cabinets of natural curiosities.

Lina Bolzoni describes how "between the sixteenth and seventeenth centuries the *ars memoriae* and collecting reciprocally influence each other, mirror each other, and exchange models and stimuli in their practice as well as at the level of theoretical systematics" (1994, 132). Just as the physician Samuel Quicchelberg attempted to design an ideal system of universal inventory making, a *theatrum sapientiae,* in his capacity as collector for the wealthy Fugger merchants and the Bavarian duke Albert V, taking Cicero as his guide,[2] so also Giulio Camillo imagined building a universal theater of the mind, a *theatro della sapienza* (ibid., 129). Memory and collection are founded on a segmented and ordered space. The cupboard or chest, important organizing receptacle for collections, serves as a metaphor for the mnemonic art. The Dominican Fra Bartolommeo di San Concordio, who authored tracts about the memory, used the symbol of a cupboard to show "that the loci of Holy Scripture are laid down firmly and in good order into the loci of his [the memorizer's] mind" (p. 137). Metaphors of the memory also come from collection spaces such as treasuries, libraries, and stockrooms. The mental place in the memory and the material place of a collection each depend on strict order among the loci, for collectibles just as for memorabilia. Whereas Quintilian's space is supposed to be as roomy and as full of variety as possible (Yates 1999, 23), since the Middle Ages utmost simplicity and regularity were sought (p. 99). The Dominican monk Romberch wrote in his treatise on the memory, *Congestorium artficiose memorie,* of 1520 that the ordering of the places was decisive. There might be many places, but they should be of average size and proportioned to the image they are to contain (Yates 1999, 101–2). He pointed out that the mental place should not be large either.

Plate 25 depicts a person measuring with his outstretched arms the appropriate amount of space needed for the mnemonic picture of a person. Although Yates presumes that this rule "evidently applies to human images, not to memory objects as images" (p. 117), the loci of the mnemonic objects shown in plate 26 are of significance as well. Here a real place with real structures is the subject of memorization—in this case it is an abbey with all the things to be found within it. The objects are presented in tabular form, as in an alphabet, and the vertical dimension of the compartments is scaled to fit the content. The objects have been assigned customized, defined spaces.

The Places of Natural Curiosities

One hundred fifty years later, discussion about suitable places and suitable objects for collection was taken up again, this time not as a fictitious locality but as the storage place of natural specimens in a curiosity cabinet. It is in

2. Cf. the dissertation by Harriet Hauger (1996) and the master's thesis by Angela Deutsch (1995).

such a locality that the physician Johann Daniel Major wanted to arrange objects from nature. In the eighteenth chapter of his book he asked "how natural objects and curios might be appropriately disposed" (1674, unpaginated). Although "containers of rarities" kept the items in proper order and classified them, it was not enough merely to make a count and use "simple alphabetical order." Many more *subdivisiones* must be made, particularly for snails and shells, which had not been afforded much attention by scientists because they "brought no bread into the house" (§ 2). Henceforth a cabinet should be "addressed in the *experimental-seculi* spirit of today" (§ 3, original emphasis), and for that, suitable "repositories" were needed to store the *naturalia.* Some collectors, as Major informs us, had their boxes equipped with standard rectangular compartments, and he "many times saw with consternation that often 2, 3, and more *species* were put into a compartment of little things" (§ 7, original emphasis). To the fixed arrangement other species could then be added only at the expense of growing disorder. For that reason he suggested installing "little open trays of tin . . . 2, 3, or 4 inches wide and 3, 4, or 5 inches long, as befits the various things (it suffices that they all have the same height and stand in a straight row inside the *Repositoriis* at the front) [to set up] therein the best specimens of the *speciei*" (§ 8, original emphasis). As an example, Major depicted a tray containing a coral specimen (plate 27). A label should be affixed to the outside, bearing the object's name. This method of arrangement had the advantage that new specimens of a species were easily added. If the objects were too small for this standard unit, the tray should be filled in with sand or the object set up inside the tray on a wax base. For oversized objects, a wood-backed slip of paper should be placed inside the appropriate tray as a substitute, so that it would be clear at least to which class the item belonged.

Although it may sound naive today to classify objects by their origins and to arrange them within a set uniform structure in total disregard of their various dimensions, the situation was quite different in the seventeenth century. The Linnaean classification, by which gentleman collectors of the subsequent century were able to arrange their *naturalia* in neat and clear order, was yet to be devised. What Major objected to was the use by many collectors of the alphabetical system. As an alternative he suggested ordering things by their proper natural classes. Simply accumulating specimens was not enough. Their arrangement, presentation, and placement both materially and within the system were essential for gaining any intellectual benefit. A collection chamber was useful to the collector only when he could pick out from among his store of objects that pertinent piece for his current research.

This tray system that Major recommended as a more transparent method of storage he called a *repositorium.* According to Zedler's *Universal-Lexicon, repositorium* means "in student quarters, a bookcase. Thus is also called every structure made up of many shelves upon which one tends to arrange in tidy order books, glasses, and sundry things" (1732–54, 31:648).[3] Accordingly, a repositorium is a system for storing physical objects in good order. Mnemonic places, on the other hand, involve a *repertorium* (Bolzoni 1994, 148), where the correlation between place and image occurs in the mind, from whence a desired combination of mnemonic places can be called up. Zedler defines *repertorium* as "a finding book or written list of particular things" (1732–54, 31:637). Stoy also referred to a "well-furnished *repertorio*" (1782, 4) in a child's memory in the context of his Academy's picture plates. But he also wanted to create well-ordered and uniform "receptacles" *(Behältnisse)* in the mind (p. 3), which the child might dip into as needed. A repositorium as a material structure of order and a repertorium as a kind of ordered index of the mind each perform the same function.[4] As a storage place they also provide material and immaterial means of orientation for the user.[5]

Presentation of places and the things they contain figures in the art of mnemonics and collecting just as it did in Stoy's didactic conception. All three areas are grounded on an empty, fictitious, compartmentalized space. All three areas use similar referencing and classificatory categories. Stoy too assigns his picture cards to specific compartments and establishes a material order for the content: his picture tableaux are divided into picture places with set localities.

The object places (box or tray, chest or shelves) were an important element of the organization of knowledge. By the end of the seventeenth century two forms of presentation became particularly prevalent—display cabinets for natural curiosities and display cabinets for merchants' wares. By the end of the next century they lost ground again in the worlds of natural history and marketing, and the materialized principle of classifying things was dematerialized and transferred to the two-dimensional sheet of paper. Leaving the realm of the three-dimensional object, representation of nature entered new media like tables, indices, and card catalogs. Only one element remains the same to this day: places and their order are an indispensable means of orientation. Loci of

3. A student album from Rostock depicts one repositorium, resembling bookshelves (see Kohfeldt and Ahrens 1919, plates 7 and 8). The dormitory furnished with such a bookcase is very neat and tidy; the other room, by contrast, is a complete mess without it.

4. *Repertorium* is late Latin for "finding location"; the Latin *reperire* means "to find." In Latin *reponere* and *repositum* mean "to replace" and "to deposit, return."

5. Meinel 1995 describes the importance of such repositories and repertories for scholars in the seventeenth century, expanding in particular on the treatment of card catalogs (pp. 179–83).

the twentieth century are just as dependent on regular order as those of Quintilian, Major, and Stoy.[6]

Cabinet Collections of Natural Curios and Merchandise Samples

The phenomenon of serious collecting, which predates modern times, was first thoroughly analyzed by Julius von Schlosser in 1908. His book on cabinets of art and curiosities of the late Renaissance, *Die Kunst- und Wunderkammern der Spätrenaissance,* describes its prehistory and distinguishes two major areas: ecclesiastical and secular cabinets. Since then an abundance of literature has appeared, differentiating more and more finely between chambers, cabinets, and magazines, their functions and forms.[7]

The collecting efforts of the sixteenth and seventeenth centuries, the period when the *Kunst-* and *Wunderkammern* first emerged, were based on a fascination with surveying the universe. The demand for completeness, which in later times was directed less toward the macrocosmos as a whole than toward individual areas, has remained. The desire to collect what is unusual, to represent parts of the outside world, united the various forms of ecclesiastical and secular chambers just as did academic and institutional specimen stores and collections by *Bürger* and aristocrat.[8] But these collections were devoted not just to a representation and symbolization of the real world, as often prevailed

---

6. Heyde 1935 and Porstmann 1928 analyze the use of card catalogs and catalog cases.

7. Schlosser's work is still a good introduction to this complex. The work by Klemm (1973) is similarly compact but less comprehensive and more geared toward scientific and technological collections and museums. Besides illustrated volumes and descriptions of individual collections (e.g., Scheicher 1979), the recent publications by Impey and MacGregor (1985) and Grote (1994) deserve special mention; the latter compiles an extensive list of authors in the history of science and art who discuss the subject of collections. Pomian (1987, 1988) is more oriented toward a social stratification of collecting, and Hooper-Greenhill 1992 discusses an interpretation according to Foucault's *les mots et les choses.* Bredekamp 1993 argues for modern significance of the *Kunstkammer* as a mediating space between art and technology. A detailed study of Ulisse Aldrovandi's collection from the seventeenth century is conducted by Findlen (1994). Finally, the recent literature on the history of collections is assessed and scrutinized by the art historian Herklotz (1994) and by Daston (1988) from the point of view of the history of science.

8. Cf. MacGregor 1994, 61 ff. During the Renaissance and early modern period, collections of art and natural curiosities were often stored together to represent the macrocosmos in the microcosm of a room. Some of these collections were called *Wunderkammer,* or cabinets of curiosities. (There was no clear distinction between *Kunstkammer,* which refers to collections of works of art, and *Wunderkammer,* cabinets of natural wonders.) The important point here is that artistic and natural objects were displayed together (see Impey and MacGregor 1985, 3).

in the *Kunstkammer* of lords and ruling monarchs, but also to analysis and classification of their objects.

The latter is most evident in the *naturalia* and store display cabinets in the waning seventeenth century and dawning eighteenth century. The structure of such a cabinet catered less to the need to depict the macrocosmos in microcosmos than to the requirement for a suitable, easily accessible form of storage satisfactory both for scientific and practical purposes. According to Pomian (1994, 112), the funds of collections can be classified into four main groups and three types of spaces: the categories of collectibles are, first, works by contemporary artists; second, art from antiquity; third, natural things; and fourth, live plants and animals. Whereas the first two groups are presented in gallery form, *naturalia* are assigned to the spatial type of a *studiolo,* and living plants and animals are tended in gardens and menageries. These forms manifest various functions for these storage facilities. The collection of Emperor Rudolf II in Prague served primarily as an exhibition of power, while that of the physician Ole Worm proposed to illustrate the natural kingdom. Collections, chambers, and cabinets were installed on the one hand to please the eye, on the other hand to advance study of nature and to function as laboratories. They had to demonstrate the good taste of their owner or promote the acquisition of new knowledge. They could serve as instruments for social advancement within the world of academe or learning, or be capital investments.

An often disregarded aspect in the literature on curiosity cabinets is the social element: these sites also served as meeting points. Virtually every exchange of correspondence between natural philosophers of the eighteenth century contains not only allusions to debates over the identification of particular objects or mentions of specimens available for exchange but also reports about encounters with colleagues at the collections.[9] The following excerpt from the Göttingen professor Johann Beckmann's Swedish travel diary about a plant swap between Linnaeus and another scholar may add insight into the "social environment" *(sozialer Raum)* that a cabinet represents: "After having picked out a few samples from among those herbs that the latter had brought back with him from America, [they] tossed all the remainder into the fire . . . , so as not to be forced to report about any of the others" (1911, 85). Just as on the market exchange, in the halls of collections intense exchange of ideas was encouraged, alliances were formed, and information was withheld. These opportunities and institutions reveal a broad range of functions and users, united

9. The changes in the late eighteenth century within museums have been noted, yet the relation between a private collection space and the exchange of knowledge has been only partially examined (see Siemer 2000).

within the cabinet of natural samples. For this reason the reference work by the physician and naturalist Michael Bernhard Valentini, *Museum Museorum,* which compiles a summary listing of collectible items and the associated literature, was designed to satisfy the quite divergent needs of its various users. Economic, aesthetic, scientific, social, and education-oriented orders overlaid each other and confined themselves within these rooms. Naturalist, emperor, and common citizen alike became managers of stores and magazines, endeavoring to create an integral whole.

Cabinets of Natural Specimens

In 1762 Johann David Köhler began a description of natural curiosity cabinets as follows: "The *Naturalien-Cabinetter* are excellent to view because they are treasure troves of the marvels of the magnificence of God. All people are inclined to observe Nature, and it is useful too" (p. 216). "All people"—that might mean scholars, who could use the collection for scientific identification of a plant, as well as interested laymen, who saw it as the cabinet of God's treasures and wonders. In the eighteenth-century mind, the primary advantage of all natural cabinets was the glimpse they offered into the natural history of the things of Nature *(Natur-Dinge):*

> For, because they [cabinets] meddle with every science of the natural bodies, neither the investigator of Nature, nor the physician, nor the mathematician can dispense with them, because they place at their disposal all sorts of foreign observations; and if a politician knows the natural properties of a land well, then he is apt to offer good counsel to make a country wealthy even without particular hardship to its subjects. (Zedler 1732–54, 23:1064)

And because the complexity of nature is not easily comprehended, the cabinet of *naturalia* with its clear order offered a valuable substitute and represented the great treasury of God on a small scale. Thus economist, physician, and apothecary meet in the same study.

Such wonderment at the multifariousness of God's work and the uses to be tapped from it was consistently founded on the conception of a perfect order reigning among all living beings and practical things. Nature's order, its three kingdoms of minerals, vegetables, and animals, was transposed to the collection room. *Naturalia* were ordered according to a rigid hierarchical scale in which each thing sprang from its immediately preceding neighbor. Man represented himself by *artificialia,* which exemplify his inventions such as instruments, machines, and elaborate carvings. A second widely used organizational method was based on the four elements, represented symbolically by natural specimens: thunderstones for fire, or shells for water. The work by Carolus Linnaeus, *Systema naturae,* which was published in 1735, had such a decisive and major impact on the approach toward collecting that existing orders for the cabinet shelves, in the storage cupboards, boxes, and drawers, were accordingly rearranged. This system, which classifies organisms hierarchically by species, genus, family, order, class, and phylum (see Jahn and Senglaub 1978), was readily adopted even when for space reasons or lack of an overview the complete structure failed to become apparent from the objects on display.

In most cases, collections of natural specimens were attached to a library, as is ideally depicted in the frontispiece for Caspar Friedrich Neickelius's book (plate 28). In the middle of the room is a table upon which objects could be placed right beside the relevant literature. This made it possible to proceed with the definition of an object in two ways. After examining the samples and consulting the appropriate works, a researcher might apply his or her own judgment, or else engage the "comparative look," comparative inspection of a number of natural specimens at once. The freedom to lay objects side by side for this purpose was gaining importance, so much so that even Linnaeus abandoned the conventional herbarium form of pressed and dried plant samples mounted on the pages of a bound volume in favor of loose specimen sheets (Jahn and Senglaub 1978, 75).

Let us take the collection of the apothecary Johann Heinrich Linck from Leipzig as an example. According to the *Index Musaei Linckiani* from 1783 to 1787, it comprised besides works of art primarily jarred reptile specimens, preserved mammals, conchs, insects, minerals, and plants. In the foreword to his index, Linck described how the entire collection was carefully set up in various rooms of his home, hung on the walls, stored in drawers, or preserved in alcohol flasks (1783–87, vol. 1). This illustrates the general practice of distributing a cabinet throughout a number of rooms in a house, since only rarely could a private scholar or pharmacist afford to devote any larger space exclusively to exhibits. According to the index, besides 800 jars and liquid preparations, there were 184 drawers containing samples from the animal kingdom, 105 drawers full of minerals, 103 of fossils, and 27 of plants. Since most objects were relatively small and light and hence easily collected in large quantities, chests, drawers, and boxes were suitable storage and constituted an essential part of the inventory of such a curiosity cabinet.

Merchandise Cabinets

Cupboards, drawers, and boxes also played a major role in the display of merchandise. These containers held various types of goods and merchandise samples and helped shopkeepers and their apprentices familiarize themselves with the variety of available products.[10]

Magazines and other forms of storage blossomed with the development of wholesale marketing. Where formerly goods were gathered at major markets and annual fairs and stockpiled, with the growth in population and resulting rise in consumption in the eighteenth century, wholesale markets developed with their warehouses. "This autonomous organization based on magazines, stocks, depots, and warehouses, . . . gradually eclipses the periodical annual markets and fairs" (Braudel 1979, 75). Wholesale markets needed such storage facilities for their bales and casks. In this system it was the wholesale distributor or merchant who had to be able to keep abreast of the growing number of products becoming increasingly available from overseas from ever greater numbers of suppliers. Wholesale merchants felt the pressure to provide a good assortment of wares, because specialty storekeepers offering only a limited selection were on the lowest rung of the professional hierarchy (see p. 332). Although warehouses already existed around the sixteenth century, with market halls serving as wholesale storehouses, public salt depositories, and granaries (see p. 77), there was a growing need for information about the strange new products beginning to appear on shop shelves. Particularly those northern European countries involved in the Atlantic trade had to familiarize themselves quickly with many exotic minerals, herbs, roots, woods, and fruit varieties. This was often difficult. Inquiries about "aldelfa" (Spanish oleander), for instance, or the value of the Dutch stiver against the Leipzig groschen could be answered only from personal experience. The trade in materials and minerals was particularly dependent on new information. The physician and spice dealer Johann Jacob Marx also stressed this: "Included in this business, true to its name, is that a materialist merchant can lay by wares many hundredfold, nay, even thousandfold, and furthermore ought and must name them by thousands of names" (1709, 3b). "In sum, such a man must be almost a living lexicon" (p. 4a). Marx also praised rarity and curiosity cabinets as a good source of information. He even recommended some worth a visit and briefly described "wonder wares" *(Wunder-Wahren)* and "monsters" (*Monstra;* pp. 7a f.). But instruction on how to acquire such comprehensive knowledge was still lacking.

10. There is as yet, to my knowledge, no thorough study on the display cabinets of merchants. For the connection between an apothecary's profession and the history of collections, see Dilg 1994.

Solutions to these open questions came from early cameralists. Concerned about the general education of tradespeople, Carl Günther Ludovici thought that a cabinet of merchandise would be useful; in it

> samples, patterns, and specimens of all goods and their various types, of all measures and weights, of all current sorts of coins from every land and town, . . . etc., should be stored and displayed, in order to be able at once to give everyone sensible concepts [*sinnliche Begriffe*] of all these things. Such a cabinet should be easy to set up in large trading towns and be open to tradesmen, as the public libraries are for scholars, for their use, to look up information about goods, measures as yet improperly known to them. (1752–56, 5:20 f.)

A description of such a cabinet of wares *(Waarencabinett)* is provided by Paul Jacob Marperger. This mercantilist, cameralist, and author of works on economics wanted to promote marketing and technological knowledge in commerce and trade and to provide an introduction to manufacturing. His goal was "to bring the economic interests of *Bürger* more strongly into play" (Lehmann 1971, 138) and to make their businesses less dependent on the state. Marperger's lexical work on the newly opened "merchant's magazine" of objects from nature and the crafts, *Das in Natur- und Kunst-Sachen Neueröffnete Kauffmanns-Magazin,* intended to provide a comprehensive education for tradesmen. Apologizing for the theoretical level at which he went about this, the author advised the reader how to most profitably engage his five senses: "Finally, still one more instruction must be advanced as to the proper and practical use of this work, that [the merchant] might have made for his study, office, or counting room a chest or cupboard partitioned into several shelves and drawers, or rather a so-called portable tradesman's magazine" (foreword). Hence a tradesman should educate himself and consult the partitioned chest as he would a reference work to find answers to his questions (plate 29). Such a furnishing is visible in the background of Stoy's illustration for this profession. The cameralist added:

> (Just as I have one for my own use, partitioned into 24 drawers, for the 24 letters of the alphabet, each of these drawers again into 64 and therefore altogether into 1,536 partitions) each one of such drawers labeled with one letter of the alphabet and those *materalia* belonging under such a name, be they made *ex mineralibus, vegetabilibus, animalibus,* or *compositis,* each collected in it according to its proper sort. (Marperger 1708, foreword)[11]

11. Marperger did not provide any information about the construction of such a storage chest. This detailed description presumably was supposed to provide the tradesman with enough of an idea to commission the work from a carpenter; upon completion he could fill it with his wares.

This was not a new idea. In the sixteenth century the Italian professor of botany and natural history Ulisse Aldrovandi owned two chests containing a total of 4,554 little drawers into which he stored his snail shells or seashells (see MacGregor 1994, 84). In this case, however, aesthetic principles predominated in the presentation of the objects. In contrast, Marperger was more interested in "bringing about unequaled knowledge concerning all sorts of merchant's wares, and if one were to place the descriptions in this book beside them, knowledge could be presented to merchants that may perhaps be unknown and unfamiliar to many merchants who have dealt with wares for a long time" (1708, foreword). He alluded to helpful reference works by which the merchant could identify his goods and samples: well-known guidebooks to collections of rarities, general works about curiosity cabinets, such as the ones by Eberhard Werner Happel and Ole Worm, in addition to a book about the famous museum of Francesco Calceolari and the *Amboinsche Rariteitkamer* by Georg Eberhard Rumpf. Marperger's instructions about cabinet design and his recommended reading list reveal how much the idea of a collection of commercial samples derived from the cabinets of arts and wonders.

It is not always possible to tell which drawers contained merchandise for sale as opposed to samples for instructive purposes. The two forms might also be combined—the frontispiece for Johann Jacob Marx's *Material-Kammer* depicts a store interior with such a drawer system (plate 30). Large drawers, which presumably are each subdivided inside, line the rear wall of the room. Barrels and bales lie about waiting to be unpacked, and a materialist can be seen at work in his vaulted space.[12] Here too a simple, not particularly decorative, unit of drawers serves the sole purpose of marking and storing the goods. These rectangular drawers represent the simplest form of accessibility of objects for daily use. Unlike the cabinets of arts, curios, or natural specimens, their main object was functionality. No piles of corals, no frills. The merchandise cabinet, right beside the *scriptor,* or writing desk, is one of the most important components of the store and constitutes the merchant's "book," whose contents were not to be read, but smelt, felt, tasted, and scrutinized. For the merchant also had to "be quick and have an answer ready, as regards the quality, samples, and sorts of any one" of the wares (Marx 1709, 3a). The quality of a root cannot be assessed by the eye alone; it must be tasted and identified by its aroma. These stores were classified alphabetically, just about the time that Major was complaining about this same classificatory dependence with respect to the more esoteric cabinets. If Marperger categorized according

12. A materialist, or dry-goods merchant, sells not only medicinal ingredients (similar to an apothecary) but also things like tobacco, coffee, and sugar.

to the natural kingdom, the alphabet provided him with the same efficient and uncomplicated access it did for his letters and books.

These modest designs with their simple indices incorporate the basic systematizing chests. And just what these rectangular containers could achieve for merchants and apothecaries alike becomes clear. Things could be safely stored yet remain readily available and in stock. Just as in botany and natural science, the ability to set samples physically beside each other for comparison was important, for only close comparison could lead to the surest and best identification. Box and tabletop together thus constitute a single semantic arena *(Bedeutungshof)* for the inspected objects. By providing a stage for the "comparative gaze," it permits classification, activates the senses, and allows haptic access. A box, then, is the ordering equivalent of the registration systems of scholars: where the latter devised methods of referencing and collection, experienced merchants, materialists, and apothecaries converted them into their "chamber of materials."

The Box and Its Semantic Arena

When regarding a merchandise cabinet as a collection, one necessarily wonders how the items were acquired and how they were made accessible. Piecemeal acquisition of *naturalia* within these private holdings, according to a set order, reflects the worldview of the period between 1630 and 1750, a worldview normally represented only in books.[13] The order assigned to the objects themselves also represents the perceived order of the world as a whole, in other words, until the end of the eighteenth century, the kingdom of God. Eberhard Werner Happel thus exclaimed in 1684, with regard to one *Kammer:* "What is all this against the matchless cabinet of masterpieces of the world's Supreme Monarch?" (1990, 223). Here form becomes the representative of the world order: Just as a *Kunstkammer* is subdivided among vaulted corridors, rooms, and "chambers," "thus one sees this grand art gallery of the world likewise magnificently and neatly partitioned." Preservation is its purpose. Much effort was expended on conserving the collection pieces. Secret recipes for alcohol solutions and embalming ingredients were devised and only rarely shared. The physician, botanist, and anatomist Frederik Ruysch, for instance, developed a comprehensive technique of specimen conservation that he was willing to divulge only for the considerable sum of fifty thousand guilders (Müller-Dietz 1989, 761).

13. See Jahn 1994 for a comprehensive discussion of this aspect.

The problem of how knowledge about herbs or other products was to be obtained or made available is not merely a matter of abstract systems of organization; "appropriation means fragmenting the world, dividing it up into finite things subject to man" (Barthes 1964, 12). Its solution was the box. Such containers formed the material basis for the eighteenth-century cabinet or established collection, which could not function without them. Boxes accompanied the collector and natural historian throughout his research, from initial gathering of specimens to final presentation. At the initiative of the Russian czar, the physician Daniel Gottlob Messerschmidt undertook an expedition to Siberia between 1720 and 1727 to explore the region and augment the royal collection of art. Many naturalists set out on such journeys, and their travel diaries often read as if their main purpose, besides the gathering of specimens and collectibles, had been to lay down a systematic web of descriptions of the countryside and its inhabitants so as to provide, like a topographical map, a multidimensional picture of the region. Economic interest in regions yet to be subsumed, new trade routes, researches on foreign languages and cultures, the discovery of rare and valuable items, and botanical studies were just some of the motivations for embarking on such adventures, which in turn became the subject of research.

According to the accounts, Messerschmidt traveled and collected during the day and started sorting and recording his samples at dusk. His travel diary reported that he used "little boxes," "crates," and "caskets" *(Kästlein, Kisten, Särglein)* to systematize the daily harvest of sample seed pods, herbs, feathers, and ores. Apart from these verbal descriptions, such containers were the only means of maintaining an overview of the objects. "27th November 1724. My *denschiks* had to fashion a sm[all] box today, 1 arshin long, 1/2 arshin wide, and 1/2 arshin tall, for separate storage of the accumulated seed pouches or *sacculos seminales*" (Messerschmidt 1962–77, 3:220). Even the construction of the boxes was deemed deserving of mention in his diary. In order to study the contents of these boxes, which he called his *labores materiae,* in the evenings, "one must have all the *serinia* [*sic*], crates, and boxes open about one . . . so as to be able to review everything wherever it be, which because of precisely this and other reasons more, cannot be done in the tent but requires completely undisrupted peace in chosen comfortable rooms and quarters" (pp. 216 f.).

The objects should be arranged within an orderly space (in boxes); the surrounding area (the room) should be clean, comfortable, and equipped with working surfaces for inspection and recording of the articles. As important as they were as a magazining tool for the traveling collector, boxlike containers were no less requisite for the cabinet. As classificatory storage chests, their design was often complex. Famous pieces of furniture like the cabinet of the Augsburg patrician Hainhofer called for more expertise than a regular carpenter could have mustered; it took a fine "cabinet maker–architect" to build it (see Schlosser 1908, 96). Removal and replacement of things inside such an elaborate design can involve complicated and time-consuming working procedures that are successfully mastered only with the aid of a guide or map. Behind the first drawer, more are hidden; others are accessible through doors elsewhere, which lead to new columns of partitions. Thus a uniform-looking unit can conceal a number of different levels (see Alfter 1986, 43). Nonetheless, they are grounded in the simple ordering principle of Major's box arrangement. In exhibition chambers, boxes and drawers were installed inside chests, stood on tables, or filled shelf-lined walls. The drawers contained compartments and were labeled inside or out. The collection articles were placed in neat rows on shelves; some lay in the partly pulled out drawers. Larger cabinet pieces like those at the museum of Michele Mercati allowed for a more specific physical classification, since each chest held a single category of rocks (plate 31). The hinged front not only protected the contents but also could be dropped down for use as a writing or working surface. As a comparison, simpler boxes can be seen in the frontispiece to the museum catalog or inventory list for Ole Worm's cabinet (plate 32). Numerous labeled boxes and trays can be seen. Lidless, they reveal the partitions fitted to accommodate their specimens: this design is the one Major censured, with a fixed number of partitions precluding further expansion. The title copperplate of Valentini's *Museum Museorum* depicts similar containers (plate 33). Hermes is represented at the lower center of the copperplate, holding a label with the inscription "East Indies." The sacks and crates lying around this ancient Greek hero suggest the long ocean journey from far-flung trading posts to this European cabinet. It looks almost like a merchant's stockroom. The bundles and barrels point to the busy trade for exotic *naturalia* and *artificialia* that merchants undertook above all in the eighteenth century (see Jahn 1994, 478).

These illustrations show the ideal cabinet or collection, with order clearly reigning among the regularly partitioned chests and boxes. Inscriptions, enumerations, and customized working surfaces for convenient inspection and identification of the holdings determine this material systematization.

Boxes, tables, and working surfaces thus occupied a major place in the cabinets and researches of scholars and merchants alike. The remaining question is when such "knowledge chests" *(Wissensschränke)* first appeared, or as the art historian George Kubler put it (1963, 39), at what point in time can their first emergence, their *prime object,* be localized. A prime object signifies the starting point of a successful solution to a particular problem. What follows are replicas, slightly modified reproductions, which themselves can appear in entirely different contexts. "In being original entities," prime objects "resist decomposition" (p. 39). Can such a prime object be identified, by which the principle of accessibility and generation of knowledge through classification and order first found widespread acceptance? We must seek the answer in the history of objects, among things that offer in such a compressed box-shaped space all the essentials of the macrocosmos of the world and the microcosmos of knowledge.

The cabinet's history starts with the printing craft and the development of movable type. The printing of books and its associated typographical culture led to the emergence of printing houses and typesetting establishments. As a consequence, collections of information could be reproduced and disseminated as never before. New professions, like that of the stampcutter (who engraved dies) or typefounder, sprang up within a new social milieu. New trade opportunities, registration methods, and the development of new working procedures and skills, specialized tools, and instruments led to a radical alteration in the way that texts were produced. The printing craft spread rapidly during the fifteenth century, once Gutenberg had published the first printed Bible in 1455–56. By the end of that century printing had already established itself in more than 250 European towns (see Giesecke 1991, 64). Presses, special paper, woodcuts, inks, movable type, and typesetters' cases were essential elements of the new medium. To these were added the precise tools and implements for the exacting skills of this new world of work: the *typographeum.* If in the fifteenth century a bookmaker incorporated within a single person all the essential steps for the production and sale of a piece of writing, specialization led to the development of several professions from the sixteenth century onward, with the consequence that from then on the tasks of a printer of books were generally reduced to typesetting, impression, and proofreading.

Whereas previous information media needed stylus and clay or wax tablet; paper, quill and ink; or a carved stamp, with the development of movable type the vehicles of knowledge were standardized, and magazine boxes became necessary accessories. Type increasingly acquired a uniform style, and its storage in cases and drawers, classified according to the letters of the alphabet, was the starting point of the world of printing. Maintaining a magazine of letters followed economic principles. Too many letters would raise costs unduly, yet enough should be on hand to allow printing of the finished pages, cleaning, and return of the letters in time to avoid having the whole process come to a grinding halt. The number of partitions in a typesetter's case decreased as the field continued to develop.

Typesetting involves picking out individual letters from the type case and arranging them in a row in a composing stick. The resulting lines of type are transferred into a galley or shallow tray until the whole column of text is finished. This column is then bound tightly together with string to prevent the individual pieces of type from falling out and is laid aside for printing. One of the earliest illustrations of a European printer depicts a typesetter at his case (plate 34). The manuscript is positioned in front of him on a stand. He is holding the composing stick in his left hand, and with his right hand he is picking out and assembling the appropriate pieces of type. This procedure was essentially the same for the eighteenth-century compositor. The letters have a fixed arrangement inside the type case (plate 35). This allows the typesetter to take up the letters quickly and to pick out the right ones with ease, though he has to be thoroughly familiar with the case and its partitions to be able to make up pages of type without error.

The typesetting procedure is thus connected to the box as a tool. Ever since this material-ordering tool came into widespread use, forming the basis for a sense-constructing result (the printed page), its purpose has not changed. Evidence can be found of containers used for food storage virtually since the beginning of the history of *Homo sapiens,* and secure storage is inherent in a box's function. Boxes or containers generally occur where things are drawn into a specific functional relationship to human beings and where human interest in an object has special significance. The box supports this significance and allows a specific field of activity to arise between hand, thing, and box. But a new development arises when the stored thing has not only a special significance of its own but also one connected to its relation to other things stored inside the same box according to a single order. This value-added significance is then used to attach new significance, new meanings to these objects outside of the box. One of the first boxes of "things" of this sort to be used systematically was the typesetter's case.

All boxes share the same function. Their content takes on the role of intermediary between the world, its objects, and man. Its underlying structural order is that of places, or loci. Both type case and cabinet as well as the Picture Academy's fields are initially empty, artificially constructed localities, which humans must fill and activate.

The analogy between the work of a typesetter and the use of a chest or box shows that the process of reading nature by means of conserved objects is not merely "visual reading" *(sehendes Lesen)* oriented toward the external traits of a plant or a mineral. A chest's classificatory partitions and the action of removal and return, with the intermediary step of juxtaposition of the objects, form a part of "constructive reading" *(machendes Lesen)* performed not just with the eyes but also with the hands. This haptic approach was an essential aspect of the eighteenth-century box as a tool and inherent in the assimilation of objects and collections. In merchandise collections it constituted a part of the process of knowledge acquisition, since it was only through direct grasping of an object that its character and quality could be properly ascertained. Curiosity and *naturalia* cabinets also were arranged in a way that the user could—albeit very carefully—pick up the objects in his or her hands.

What Aby M. Warburg related of the artistic process also applies to the eighteenth-century cabinet: exploration of the object "by the groping hand" is "equidistant from imaginary grabbing and conceptual contemplation" (1992, 171; cf. Gombrich 1970, 290). This also corresponds to Stoy's didactic approach with respect to the *ars memoriae.* For between imaginary apprehension in the memory, the dedicated memory place, and conceptual viewing of the written or verbally formulated results of the natural philosopher lies haptic exploration of the material object. "A pun of Warburg's—*Greifen – Ergreifen – Begreifen – Ergriffenheit* [grab → grip → grasp → gripping]—indicates the symbolic role of the hand in the development of rationality. The stimulus is the 'grabbiness' impulse that—like a child—wants to take possession of everything material that comes within reach" (Barta-Fliedl and Geissmar 1992, 61).[14]

The historical phase of merchandise and natural cabinets is significant because at this stage conceptual viewing and the boxed object meet within their ordered categories. The recording system of merchant or naturalist (indices, alphabetical order, arrangements according to specific classifications) was still identical with the material location system (drawers, compartments). This intersection between written and material location systems fell apart only toward the end of the eighteenth century. Before then, however, pedagogical theorists, private tutors, and publishers discovered the box for their work.

---

14. Leroi-Gourhan (1964–65, vol. 2, chaps. 1 and 2) attaches a more than symbolic role to the hand in the development of rationality, assigning it an action-oriented, materially directed function.

Tableau C

Plate 25. Human image on a memory locus (1533)
(Yates 1999, figure 3, p. 118)

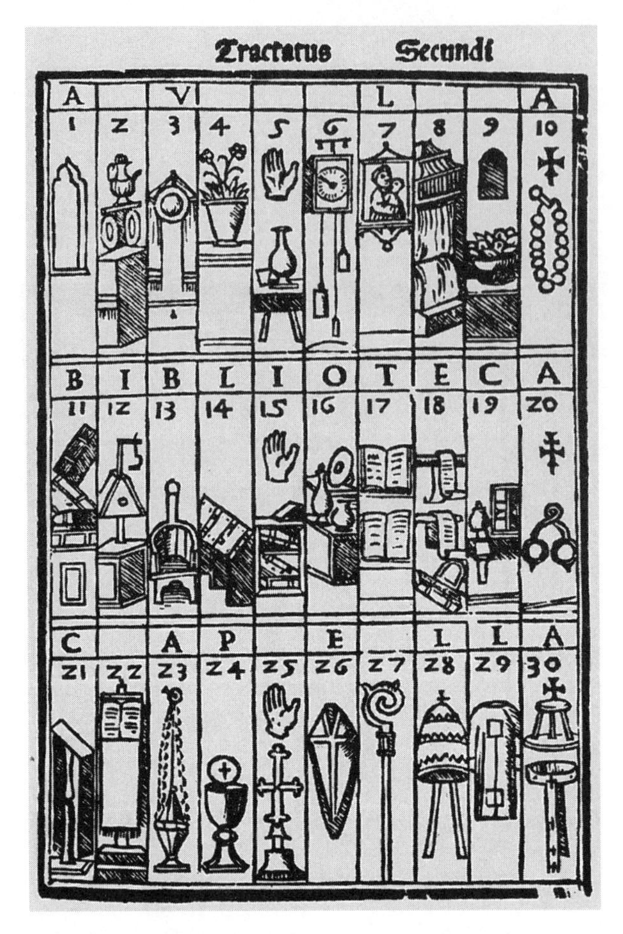

Plate 26. Memorization objects (1520)
(Yates 1999, plate 5b, facing p. 112)

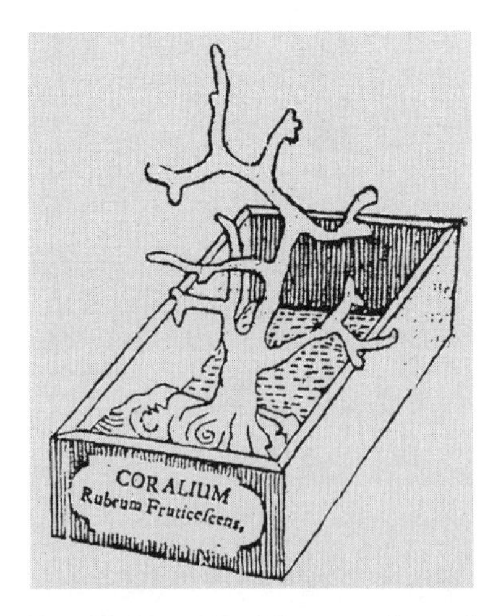

Plate 27. Tray containing a coral sample
for a cabinet (Major 1674)

Plate 28. Frontispiece for the *Museographia* (Neickelius 1727)

Plate 29. The merchant in his chamber (Stoy 1780–84, copperplate 11, field 6)

Plate 30. Frontispiece for the *Material-Kammer* (Marx 1709)

Plate 31. Cabinet piece for Mercati (detail, 1717)
(Impey and MacGregor 1985, figure 2)

Plate 32. Frontispiece for the museum catalog of Ole Worm (1655)
(Impey and MacGregor 1985, figure 51)

Plate 33. Frontispiece for *Museum Museorum* (detail, Valentini 1704)

Plate 34. A typesetter and printer in a dance of death (detail, 1499–1500)
(Bogeng 1930–41, vol. 2)

Plate 35. Front view of a French type case (early eighteenth century)
(Barthes 1964, 24)

6

THE BOX AND ITS USES

By the second half of the eighteenth century, cabinets and collections could no longer pose as an "integral unifying space" (*integraler Einheitsraum,* Braungart 1989, 118). Systematic, comprehensive storage of objects was no longer possible, and very different exhibition depositories emerged, specializing in particular topics: natural cabinets, which later evolved into natural history museums and galleries or art museums. They too were subject to a rigid order but did not strive after universal representation of the macrocosmos.

Nevertheless, the "unifying space" did not quite disappear. It remained alive in the belief that a universal basis for knowledge existed that must be passed on to the learning child. In this regard elementary works like Stoy's or Basedow's continued the project begun by Comenius, a guided promenade through the world *(Weltbegehung).* "Let me guide you through all things; I will show you everything; I will name you everything" (Comenius 1685, introduction). By "all things," Comenius meant heaven and earth, animals and plants, professions and objects, which concept the eighteenth-century outline found no reason to argue with, although its order had become more complex and detailed.

Links between the Mnemonic Art, Cabinets, and Educational Concepts

The three famous accounts from the seventeenth century describing utopian states and cities, Bacon's *New Atlantis,* Andreä's *Christianopolis,* and Campanella's *Civitas solis,* are based on a common notion. They reflect a perfect world in which all the essentials are preserved and in which education of the younger generation plays a major role. Ranks of pictures line the walls and streets; rooms are equipped with models and collections of natural specimens, arranged in separate departments of knowledge. Walking through the streets, one would gain insight only by conceiving all impressions of the city as a complex interconnection of knowledge, as a single mnemonic space. This type of "global review" *(Weltzusammenschau)* constitutes what Bredekamp calls the "cognitive form of the *Kunstkammer*" (*Erkenntnisform der Kunstkammer,*

1993, 68). Here an ultimate system of the natural and philosophical sciences, replete with laboratories and demonstration rooms, corresponds to the image of the utopian state.

From the seventeenth century onward, pictures and objects gathered increasing importance in children's education. Johann Amos Comenius, who considered himself a pupil of Andreä, and the pedagogue Wolfgang Ratke put the res and pictures on the Occidental curriculum (see Dolch 1982, 283 ff.). The emerging link between a spatial chamber and pictorial presentation of knowledge was not so much explicitly formulated in the children's literature of the turn of the seventeenth to eighteenth century as it was actually put into practice. These visualizations provide a link to the Picture Academy and throw light on its realization.

The copperplate engraving entitled "Pleasant Diversions with Pictures: Thus Fondly Arranged for the Amusement of the Young" (plate 36) is the frontispiece of a book of the same title *(Angenehme Bilder-Lust: Der lieben Jugend zur Ergötzung also eingerichtet),* which appeared in 1760 via the Nuremberg publisher Peter Conrad Monath, famed for his children's books and adolescents' literature. Inside a room opening out to the viewer in central perspective, an adult—presumably the children's preceptor—is welcoming a second man, who seems to have just entered, still with hat and épée. Eighteen children—their attire suggests that they belong to the nobility or wealthy *Bürgertum*—are wandering about in small groups. The walls are completely lined with pictures separated from one another only by very narrow strips.[1] The children are left to look at the pictures on their own and are pointing out to each other the depicted figures, animals, and scenes. They seem to be describing what they see. This picture chamber created "for the amusement of the young," the gesticulating children in particular, and the picture arrangement all suggest the visualization of a mnemonic place.

In imitation of other children's books and illustrated volumes like the "Picture Hall" *(Bilder-Saal),* "Picture School" *(Bilder-Schule),* and "Picture Cabinet" *(Bilder-Cabinet),* this copperplate frontispiece provides a spatial interpretation of the principle of visualization by assemblages of images, as in a gallery or museum. The *imagines agentes* find their loci on the wall of the baroque chamber that the children walk through. The title copperplate for the *Historischer Bilder-Saal* from 1697 (plate 37) is similarly designed. A child

---

1. As has already been seen in the frontispiece for Voit's *Schauplatz* (see plate 2), the environs of this classroom scene are depicted as a garden. The baroque interior is enveloped in subdued light, in contrast to the sunlit natural world outside, which casts distinct shadows through the open doorways.

is being led by two familiar figures, Clio and Minerva, through the gallery, which is covered from floor to ceiling with pictures.[2] This symbolic scene of instruction, engraved seventy years before the appearance of the *Angenehme Bilder-Lust,* contains some clues to suggest that the origin of galleries is the cabinet of art and wonders. In the foreground is a table on which a small cabinet is standing open. One of its drawers is pulled out to reveal small objects, probably gems or cut stones. The busts of famous men stand sentinel on the top of the cabinet and embellish the walls on their ornate pedestals. Clio also introduces the Picture Academy as a gallerylike exhibition of pictures (plate 1). Hanging on the wall are memorable "active images," striking depictions like Cain's assault on Abel. These are what the children are supposed to look at while Clio keeps the history book safely on her lap.

The idea that these three frontispieces could well involve mnemonic places is supported by one of the engravings in the ninth field of the Picture Academy (plate 38). A boy appears in a self-absorbed pose: his back slightly hunched over, his left arm hanging down limply, and his right hand touching his forehead. The child is presented from the side, standing before a wall of pictures. The wall is subdivided into a number of fields illustrating the boy's own tale as a liar, as the Academy's first text volume informs us (Stoy 1780–84, 116–18). From the series of images and the text it is clear that this is a moment of repentance when the boy resolves to mend his ways. The pictures on the wall beside him reveal his introspective review of his mischievous deeds. These memories of events are depicted sequentially and represent the main stages in remarkable, memorable images.

The representational form of these copperplates is based on the Aristotelian notion of the memory (see Florey 1993, 157 ff.). Accordingly, the imprint of a perception by the bodily senses activates a mental image. This image is, as it were, a copy of the remembered object or event. Through this imprint and its image—Plato uses the metaphor of a wax seal—the past (event) is connected with the present and makes memory possible. Memory means, in this context, the activation of images.[3]

This connection between image presentation, mnemonics, and children's instruction within a room finds its parallels already in the sixteenth century. Utopian city-states and copperplates are not the only evidence of the idea of

2. Meier (1994, 24) identifies the female figure holding Mercury's caduceus as Athena. However, the pair Clio and Minerva is more likely in the context of education (see also Stoy's frontispiece, plate 1) and the symbolisms of lexica and encyclopedias (cf. the frontispiece in Halle 1761–79, vol. 1).

3. These notions still applied in the eighteenth century, and physiologists like Albrecht von Haller adopted Aristotle's formulation of the memory as their own (Florey 1993, 164).

teaching through images and of referencing in association with the chamber. A gallery with paintings even on the ceiling exists as the famous *studiolo* of Francesco de Medici in the Palazzo Vecchio in Florence (plate 39). This ornate chamber conceals cabinets filled with *naturalia* and *artificialia* behind its allegorical paintings from mythology. No labels indicate the contents of each cabinet; the paintings themselves provide the clue (see MacGregor 1994, 65; Braungart 1989, 114 ff.). Thus art refers to nature. These two fields confront each other as human artifact versus work of Nature. The mnemonic *imagines agentes* harbor the knowledge of the world; the paintings of the Medici *studiolo* reiterate the cabinet contents. They can be described as materialized memory signs.

Another picture comparison, relevant to seventeenth-century pedagogy, supports this intersection between chamber, memory, and instruction (plate 40). The frequently reproduced illustration of the court library and *Kunstkammer* in Vienna shows the same custom of promenading through the room that is seen in the *Angenehme Bilder-Lust* (plate 36). In both illustrations the visitors are in pairs, engrossed in conversation.[4] While the adults are poring over the contents of open drawers and cabinets or are looking up at the wall hangings, the children—imitating the adults by their civilized attitudes—are standing in a picture library designed just for them.

In drawing this parallel between chamber and educational form, another parallel in content is provided by a comparison of the Academy's dedication plate with similar depictions of research utensils such as the globe, compasses, protractor, and painter's palette. An equivalent selection of objects appears in a vignette of Halle's *Werkstäte* (1761–79, title vignette) as well as in one for the records of 1728 by the Instituto delle Scienze founded in Bologna in 1712 (see Bredekamp 1993, 56). All three cases involve exemplary objects from the three kingdoms of nature, from antiquity, and from the arts and the crafts. Their heterogeneous arrangement points to two things: their functions in academic research and the commercial applications developed out of the resulting scientific findings.[5]

Just as the Instituto delle Scienze originated from a cabinet of arts, the Academy drew upon the same traditional ensemble of objects for its frontispiece and thereby signals their common heritage.

---

4. The arrangement of living creatures in couples is an essential iconographic motif of Noah's ark, with which the *Kunstkammer* is often equated (see Neickelius 1727, 9; Köhler 1762, 217 f.).

5. These utensils can be described as a typical *Kunstkammer* ensemble. Horst Bredekamp (1982, 546 ff.) has shown that there is no essential contradiction between the cabinet of arts and the commercial use of the things they contain.

Three basic and common features identify this historical conflation of Stoy's Academy with mnemonic art and the cabinet phenomenon: (1) In *ars memoriae,* just as in *Kunstkammer* and children's educational forms, knowledge is presented spatially. (2) This presentation depends on a parceling and ordering of the world, which is achieved by assigning a specific locality to every single object. (3) These models call for encyclopedic totality, constricting the world within a miniaturized, representative space.

But above all, the models are grounded in a rigid order that must be adhered to in all circumstances. This order provides orientation and makes apprehension of the world possible. In eighteenth-century education it was important that "just the simpler things be done with the children at first, the more difficult things, though, as their aptitude grows" (Stoy 1780–84, directions, 14). In this ordering system everything had not only its proper place but also its proper class. The ordering of human knowledge was therefore oriented not least by the Cartesian rules of logic. Descartes explains this step of mental classification in the third of his "rules for the direction of the mind": "to conduct my thoughts in order beginning with the simplest and easiest to understand, and advancing little by little, as by degrees to the knowledge of the most composite; and supposing order even among those that do not naturally precede one another" (1964–74, 6:18–19; cf. Garber 1992, 44). "Supposing order," that is, arranging things in an ordered sequence that engenders insight and so could—as Johann Heinrich Alsted put it in 1630—"breathe life into them," numbered among the grand programs undertaken in the seventeenth century.

Order: The Basis of All Knowledge

The concept of order acquired a more practical and worldly flavor in the eighteenth century. Pedagogues, preceptors, and publishers came to value it as an educational approach. The prevailing catchwords defining the discussion around children's education—emulation, civility, and measured conduct—were joined by those of establishing, maintaining, and preserving order and evolved into a central aim of educators.

The first part of Johann Caspar Lavater's "Rules for Children" (*Regeln für Kinder,* 1794) is structured around individual captions that reiterate the rules of conduct; the second part combines these rules within clear exemplary stories. The key concept of order reappears under the captions "God," "gratitude," "truthfulness," and "fairness," and likewise under "stubbornness," "bad mood," "pride," and "slander." Lavater's rules illuminate the value of order as an indispensable part of life:

> Order, my child, how can I recommend this to you enough? You should know, or easily be able to recall, where each of your objects—or those that have been entrusted into your charge—can be found. All your things ought to be arranged and ordered so that you can survey them easily and produce and use them at any hour without great effort. Habituate yourself as early and as strictly as possible to orderliness. Through order you spare yourself countless vexations and most unpleasant agitations. You savor life doubly by order, and, by disorder, you throw a great part of your life away. (p. 41)

Accordingly, order activates the memory, saves time, and serves as a valuable basis for an organized, directed lifestyle. Thirty years earlier Rousseau had already noted the importance of even young children's respecting the rules of order: "At the beginning of life, when memory and imagination are still inactive, a child notices only what momentarily affects his senses. His sensations being the first material of his knowledge, offering them to him in a convenient order prepares his memory for furnishing them to his understanding one day in the same order" (1969, book 1, p. 120). This support of Lavater's statement concurs with the virtue of clear structural form as offered by Basedow, Villaume, and Stoy.

In his guidebook to parents, *Das Methodenbuch für Väter und Mütter,* Basedow stated that order could not be evaded, even calling it the foundation of modest prosperity: "Love of order and neatness alleviates virtually all homely and civil duties. From the first years on, you should oblige your children to keep everything that has been placed in their charge among the others in the right place and on the right shelf; and after finishing the business or game, to put everything back in its proper place" (1770, 178). So everything had its place. Things were locked into their assigned spot and always had to be returned there to maintain a clear and sensible sequence.

Anchoring order in the mind is the goal of one children's game conceived by the pedagogue Peter Villaume: "The game of swift and orderly arrangement has a thousand variations. E.g., everyone is given a pack of playing cards messily laid, or some books or other things" for the child to return "to a prescribed order" (1787, 391). Whoever had to be corrected was the loser of the game. This game of "orderly arrangement" was intended to provide the groundwork for able, tidy behavior.

The places of things played a special part in inculcating this positive quality. The sense and purpose of order is to be able to recall with ease where each object can be found. Basedow asserted that it was a simple matter to retain and repeat words but that objects were much more difficult. That is why children should be accustomed "to noticing so exactly all that they see and hear in particular places and under particular circumstances that no incident of any significance is left out and the order of the things and events stays intact" (1770, 212). He suggested starting first with easy tasks, for example, remembering one object. Then the children were to be told to try to describe a whole room, and finally a "large area" (ibid.). The path from object to room and eventually to an entire landscape follows the step-by-step advance and expansion of a child's horizon.

To reach a similar end, Stoy drew up his tableaux and box. The Academy's foreword reads:

> By this arrangement, compilation, and connection between so many subjects, I sought to enrich the minds and hearts of the young early on with immeasurable treasures, and in time to make them familiar with the *spirit of order* [*Geist der Ordnung*], which, as is generally ascertained, infinitely simplifies learning as well as retention and application of what is learned. (p. 11, original emphasis)

Such representations of values contained in these directions on how to exercise the memory and condition tidiness followed the general consensus about raising orderly children. What is also interesting here is the allusion to possession. Order was the order of a possessor; a child should always know where all his or her things are (see Lavater 1794, 41). And a very specific type of order was involved: things must be "arranged and ordered so that you can survey them easily" (ibid.). "From the first years on, you should oblige your children to keep everything that has been placed in their charge among the others in the right place and on the right shelf" (Basedow 1770, 178). One should not excuse the children "when they cannot find or lose such things as have been entrusted to them for safekeeping or even for locking away" (ibid.).

Order is, above all, an exercise in value management. A scrap of knowledge in the mind, an exquisitely painted teacup or florin coin in its box must not disappear. This emphasis on order casts light on another observance of composed structures, namely social station or class. The values that eighteenth-century educators stressed so much should be regarded in the context of the *Bürgertum*'s gradual gain on the privileged classes, without there being any intent to overturn the social order. Hard work and moral ethics, diligence, and neatness were the constants of this developing middle-class mentality.[6] One rudimen-

---

6. Paul Münch (1988, 66–71) suggests order as a basic value of the hierarchical society of early modern times and points to its transformation from one to many orders by the end of that century.

tary exercise of this, which has hitherto received little attention in the related literature and hence will be focused upon here, is collecting. Collecting was not explicitly called a virtue but was nevertheless the means toward acquiring knowledge. Collecting leads to possession, and everyone, be they apothecary, simple tradesman, or mere child, could start a collection. The pedagogue Ernst Christian Trapp (1787, 39) gave a synopsis of the intellectual value of collecting and thereby aptly characterized this typical eighteenth-century pastime as the material or variation of form. Order was the form that generated certainty and clarity.

The Virtue of Collecting

Collecting delivers; it provides the content or "material of form" (Trapp), or the basis of a given order (Basedow). For educators, it offered in the first place an occupation, an ever useful working game or playful job for children.

In describing eighteenth-century collecting as a virtue, I do not primarily mean moral virtue or the observance of ethical values so much as the more basic idea of usefulness or "soundness" (*Tüchtigsein;* cf. Grimm and Grimm 1984, 22:1561).[7] Understood from this point of view, collecting acquires greater importance particularly in that century, with its heavy emphasis on virtue.

The word *collecting* was a normal part of the vocabulary of any eighteenth-century writer. It was the most frequently used term to legitimize a text and the effort expended on it: Stoy reported his "four-year long collection, incorporation, and revision" (1779, 3) of the material for his Picture Academy.[8] In Göttingen a magazine called "The Collector" *(Der Sammler)* contained town news as well as literature samples and reviews. In the foreword to its issue for 1736 the editor almost apologetically alluded to his labors over many years in collecting the material and emphasized that he was merely drawing from this existing store, for which reason he could earn neither praise nor censure for the writings, as he was but the compiler.

No criteria concerning quality seem to have been applied toward these collections, since less interesting contributions appear among them—and the

7. Dressen (1982) considers the fostering of diligence *(Verfleißung)* as an educational process in the eighteenth and nineteenth centuries; compare Münch 1984, with a documentary compilation of *bürgerliche* virtues.

8. Findlen (1989, 63–66) has shown that collecting as a theme did not first come into focus in the eighteenth century and is the basis of all learning. She characterizes the compilational and collational efforts of scholars as the "encyclopedic strategies" of the Renaissance.

authors and editors hastened to apologize for them in advance. Johann Heinrich Gottlieb Heusinger relegated collecting to the very bottom of the hierarchy of cognitive activities. This philosopher, pedagogue, and later auctioneer of books and coins subdivided human intellectual powers among four categories: "creating, collecting, modifying, and implementing" (1794, 70). For Heusinger creativity stems from the powers of sensation *(Sinnlichkeit),* the mind *(Verstand),* and reason *(Vernunft);* the modifying category combines these first two powers; and the implementing category refers to opinion and judgment. The collecting power, finally, consists of "that which retains what is acquired by sensation, the mind, and reason and which places us in a position to become conscious of this store, if we so wish. Here is where the memory belongs" (p. 73). Besides his choice of words, the relation Heusinger drew between the implementing and collecting powers is also interesting. The latter are the reservoir that implementing power can tap in preparing the way for human understanding. The so-called sensual relationship between external stimuli and internal impressions points to a changed interpretation of instruction. In this respect Trapp criticized overly complicated teaching of "superficial" knowledge. Instead, elementary knowledge, which a child probably is already partly acquainted with, should first be entrenched. "As though it were not the natural way first to gather what is lying on top before breaking through the surface; as though one should not have first taken up the gold nuggets lying about before going down into the shafts" (Trapp 1787, 73). Although collecting is but a handmaid of the mind, yet—as is repeatedly stressed—neither education nor the exchange and advancement of knowledge was possible without it.

Also implied in this occupation, according to the eighteenth-century canon of values, were the commendable qualities of persistence, continuity, and will or ambition. A collector won social prestige, became a cosmopolitan within the small, specialized, and limited world of his own collection, and knew how to handle the most obtuse things with the confident hand of a connoisseur. Collecting was regarded as an excellent proof of one's industry and a good remedy for sloth. This anthropological notion of the busy-as-a-bee assiduous citizen[9] agrees with the following reflection by the theorist of curiosity cabinets, Neickelius: " Human nature has among many other splendid qualities also primarily this one, that it never can be idle, but must be devising something at all times, while our senses and minds are ever in constant motion" (1727, foreword).

9. Such metaphors for the activity of collecting also appear in eighteenth-century education; see te Heesen 1996.

Occupation

Collecting is an endlessly repetitive and never idle action and an essential element of Stoy's design. Whereas in tableau form the Picture Academy stays booklike, at best separable into individual picture plates, the picture box constitutes a fundamental change. Ordering the picture cards into their compartments gives easy access to specific subjects where, previously, the structural sense could be established only from the central vantage point of the copperplate's religious field. The box allows one to construct a sensible system of one's own; the cards can be grasped, looked at individually, and reassembled in a new order. The basic harmony or rational unity of the tableau has been cut apart into a collection of loose slips, eliminating the necessity for guidance from one piece to the next. Instead, a pupil might simply look at the individual pictures on his own or even compose new tableaux, making the most of the variability of the cards. In a box, the world order has been brought into a form that fundamentally permits expansion of the individual divisions (fields); new bits of information are easily incorporated.

Moreover, whereas the tableau engages the eye alone, here order may be interpreted as a physical exercise. From this point of view the box starts to function only with the child's manual grasping of each card. By their very nature, cards invite you to take them out and refile them again, to mix and match them for the sheer fun of it. Thus, in this form the Academy has to be considered in the context of games or entertaining employment. With the literarization of children's knowledge came the attribution of content to objects per se, or to manipulable realia, with the aim of nurturing and training a child's intellectual proficiency. The Picture Academy implemented such recently elevated realia. With his entertaining cards, Stoy provided a "first playful instruction in pictures" (1782, 1).

Game and lesson, enjoyable instruction and work, however, are scarcely distinguishable in the approach advocated by Stoy and the Philanthropinists. Play was understood as a voluntary activity, as a source of fun and joy. To what a limited extent this actually agreed with the games peddled by educational institutions and books is revealed in numerous accounts. It was more a matter of the potential link between enjoyment and practical or instructive effect. Play was a means toward an end and primarily had to promote the children's experience and physical development. Their bodies should be conditioned for labor, their senses sharpened. This modern engagement of real things in a child's education boiled down to the educational practicality of raising a good citizen of the state.

Games in the classroom included role playing (such as acting out a manufacturing process), activities that train and develop the senses, and memory exercises. One memorization game by Joachim Heinrich Campe was intended as a review of material just learned in class: "The teacher draws a square on the board, with 6, 8, 9, or more fields, according to the ages and abilities of the children, and writes in each field one of the words to be retained that arose in the past lesson. When all have looked for a while at the fields in silence, the teacher turns the board over" (1778, 406 f.). The children were supposed to memorize the words with the aid of the numbered fields. Stoy too emphasized that only initially was the Academy intended as a child's game, "as a pastime for hours of leisure . . . until, gradually, he grows up for manly work and for a sort of elaboration" (1780–84, sample, 6); and ultimately it ought to continue to be of use throughout his whole adult life (1782, 3). The goal of all these efforts was summarized by Ernst Christian Trapp: a child habituated by this play method to work "shall become industrious and, accordingly, shall henceforth willingly take on prescribed tasks" (1787, 123).

Thus Stoy and the Philanthropinists were talking less about games than about tools for teachers and instructors. They steered a child's behavior by conveying the class material in an enjoyable manner and by providing variety in an otherwise dry lesson. Finding suitable playing and teaching material was not simple. All attempts to combine usefulness with pleasure center around a pupil's consent—and involve less a support than a surreptitious stratagem. Trapp expressed it this way:

> Children like nothing better, now, than play; i.e., they enjoy being employed in a manner that sets the body in motion, does not tax their minds with too much unpleasant exertion, easily calls up already existing ideas, and unwittingly extends or adds new ones to them; whereby the cycle of modifications and variations, which albeit often may be but a few, are unvaryingly repeated many times over, and whereby there is no compulsion other than what they place upon themselves, because it is part of the game, no exertion other than what they freely do; and this compulsion, this exertion is in some games not little; and the concentration required in some, quite much. (p. 106)

The boxed Academy was one possible solution. While playing with it a child could move about without too much effort and at the same time might add to and strengthen his or her knowledge; the many pictures provided plenty of variety. What games—besides the Picture Academy—did eighteenth-century educators have in mind when it came to imbuing and reinforcing the virtues of order and collecting in the next generation?[10] Games like those mentioned in the cameralist Johann Beckmann's travel account when he described the blind son of a merchant: "He spent his time making . . . out of many hundreds of playing cards boxes with many drawers and compartments, whose regular construction was astonishing" (1911, 84).

Toys first began to be mass-produced in the eighteenth century (see Berg 1983, 744). Although at that time Rousseau was already questioning the value of prefabricated toys for children, teachers and educators were still drawn to them in their search for aids in training intellectual and physical abilities in a "pleasantly done" lesson (Stoy 1780–84, directions, 2). Reiner Wild attributes growth in the availability of toys to the societal development that was "isolating the younger generation from the world of work": "This is also one of the reasons for the enormous expansion in the area of schooling since the 18th century" (1990, 57, 58). The material made available to children also indicates this development. Besides introducing a particular sphere of the working world, science and technology, toys also took on a representational function. "In modern times, much more strongly than in the Middle Ages, children's games came to be regarded among noble and wealthy circles as leading to behavior suited to their station. That is why a bought toy always had a representational function, besides its function as a plaything. Its external form represented an excerpt of that world into which the child was supposed to be introduced" (Retter 1979, 67).

At the end of the eighteenth century, conventional toys of the middle class included, besides dolls, hobby horses, and lead soldiers, little general stores and trinket shops, spice stalls, and pharmacies (p. 65).[11] A catalog by Georg Hieronimus Bestelmeier lists a shopkeeper's stockroom with a separate accounting cubicle (plate 41). The shelves are well stocked with barrels, sugar loaves, bales, and small crates. There is even a scale with weights, and a writing desk can be seen in the office (Junker and Stille 1984, 86). As in a real shop, every product has its place on the shelves. Besides such detailed specialty toys (see also Gütle 1790 and 1791), there were containers of every description designed specifically for children. The manufactured wares Johann

10. There is no basic literature on historical or current applications of collecting in children's education; the papers by Duncker (1990), Fatke and Flitner (1983), and Schloz (1986) deserve mention. Although not written for a pedagogical readership, Cahn 1987 contains valuable ideas about possible motivations for collecting; likewise, from a psychoanalytical point of view, Muensterberger 1994.

11. The fashion of miniaturized workplaces and milieus predated the eighteenth century. Ariès points out: "This taste for representing in miniature the people and things of daily life . . . resulted in an art and industry designed as much to satisfy adults as to amuse children" (1962, 69).

Ferdinand Roth described include typical Nuremberg toys: "Little chests of drawers, painted, for children in many [model] numbers" (1800–1802, 2:298). These various designs were used as jewelry boxes or storage for the avid young collector. Besides paint boxes and crayons, dolls and tin figures, there were little household effects, wooden or tin fruits for the miniature shop displays, little lottery and bingo boxes, and optical picture boxes (pp. 304–14). The variety of card games resembling those offered in Weigel's art store were, according to Stoy, a "sensible pastime, [for] putting words together in many different ways," and "each of these collections, just as many as there are sheets of cards, [are stored] in fine little chests, with directions as to use of the same" (1780–84, sample, 8). A chest for the letters and syllables of such a game is depicted in the frontispiece for a French children's textbook (plate 42). Two instructors are standing by this card catalog, and four children from the French royal family are busy sorting, refiling, or forming words out of the separate letter cards (they might later on work in a post office).

As if commenting on this illustration, Georg Christoph Lichtenberg wrote in one of his scrap notebooks: "Hence I should advise all young people to order all new words neatly into their classes just as one does minerals, so that they can be easily recovered when asked for or when wanting to use them oneself. This is called economy of words and is just as profitable to the mind as economy of money is to the purse" (1973–75, 1:86). This profit rests on the maintenance of order. Just as with the Picture Academy, here too the game revolves around the action of pulling and refiling cards. Although the toys might very well stimulate an entirely different course, leading to a mess or a quite different game, the pedagogues anticipated and forestalled this by advising that the Academy be used only under the instructor's supervision.

One important teaching aid of the eighteenth century was "rarely read, even though it lies before every one's eyes" (Pluche 1733–39, 1:v). Right next to the Book of Revelation is the Book of Nature, hence the natural kingdom, which displays before the rational mind's eye the three main classes of plants, animals, and minerals, in demonstration of God's magnificence and omnipotence. Studying nature was—as children's books, weekly periodicals, and calendars repeatedly extolled—devout practice of the faith.[12] It is in this capacity that it entered the curriculum. Even before its gradual emergence in Philanthropinist-oriented schools, cabinets of natural specimens such as the one maintained by the cameralist Julius Bernhard von Rohr formed one element of elite ed-

12. This physicotheological motivation is analyzed in depth in the study by Philipp (1957). See also Krolzik 1980.

ucation: "It would be desirable if, to the libraries of great lords, which are open to all for public use, were added certain natural cabinets in which one could find gathered within a single *compendio,* according to the latest standards, what one should otherwise have to seek scattered throughout the world" (1718, 464). Then "at particular hours of the day, young people" could attend class there (p. 465). Clearly they could not acquire all the essentials from mere speculations and "verbiage" (*Wörter-Kram,* p. 145). Unmitigated examination of natural things ought to be instituted, so that learning would be achieved not primarily out of books but—as Bacon would have it—firsthand from the things themselves. Comenius, Ratke, and Francke also postulated such an empirical basis for the study of nature. They advocated the construction and use of cabinets of *naturalia* and realia. One of the first school cabinets was set up in the Francke Foundation orphanage in Halle (see Storz 1962 and Müller 1995). From its inception, merchandise samples and models of human artifice were included in it, forming a meeting point between nature and technology. Such cabinets may be regarded as successors to the *Kunst-* and *Wunderkammer.*

The establishment of school cabinets ran parallel to the emergence of the *Realschule,* a more technically oriented form of secondary school. At the beginning of the eighteenth century the preacher and schoolmaster Christoph Semler introduced the first school of its kind, "in which [the material] is demonstrated more directly and explained in every detail" (cited in Maassen 1960, 25). To such a program, which comfortably incorporated a woodworking lathe, a chemical laboratory, or optical instruments, a natural cabinet was necessary. Such instruction—Semler continued—was crucial for the welfare of the state.

Such applications of the cabinet only became more widespread and numerous with the Philanthropinist movement, however (see Stach 1972, 23). It advocated exposing children to daily contact with physical objects, in order to stimulate their senses and stir their zeal. Johann Bernhard Basedow's *Naturalienkabinett* was divided into nine sections. Four were devoted to the natural kingdom, two were stocked with mathematical and technical implements, one was filled with a set of copperplates depicting historical scenes, and one contained familiar household objects. The last section displayed merchandise samples (pp. 25–26). Basedow's instructions, which appeared in 1770, exemplify the Philanthropinist aim as formulated by Ernst Christian Trapp: "teach him [the pupil] to notice everything, perceive everything, visually recognize everything" (Trapp 1787, 144). Accordingly, physical objects were more highly placed than copperplates, which nonetheless offered better material for instruction than any verbal explanation of objects.

The requirements of intellectually stimulating excursions were transferred to the school cabinet as well. This cultivating tradition, which hitherto had been limited to young noblemen (see also Rohr 1718), found its equivalent here in middle-class circles. Numerous guides like "The Travels of Salzmann's Pupils" *(Reisen der Salzmannischen Zöglinge)* appeared. These aids to appropriate and profitable visits of natural cabinets and collections indicate the instructive purpose of their subject matter (see Köhler 1762; Rudolph 1766). On these excursions pupils were encouraged to start a collection, and their active participation was deemed more highly than previously. The "virtue of collecting" *(Tugend des Sammelns)* was even specified, for instance, in Heusinger's curriculum (Teuscher 1911–12, 174).

The pedagogue Christian Gotthilf Salzmann took his pupils to libraries, monasteries, anatomical theaters, and collections. Concerning one destination, the natural cabinet at the castle in Jena, he reported the children gaping "wide-eyed at the sight of a stuffed camel and walrus, which latter they inspected with particular attention" (1793, 171). According to Salzmann, only upon careful examination might nature be understood. These insights were strengthened when each pupil started his own collection of natural specimens and gradually extended it. Salzmann's source was his own personal experience. As a young student he was familiar only with ancient languages. He knew absolutely nothing about natural history. "But then I encountered among my classmates a fellow countryman who was very avidly collecting insects. This avidity soon infected me too. I set out to look for caterpillars and butterflies" (p. 161). Ultimately, his collection had a "very remarkable influence" on his "whole way of thinking" and confirmed his view that a child's education should follow this tack. The naturalist and scholar Franz von Paula Schrank made a similar confession in his autobiography of 1766.

> [I] too promised to start a collection, namely, of the most hideous among the insects, spiders. Thenceforth not a spider was safe from me anymore. I skewered every one of them as they came and pinned them to a soft wooden board. When the board was full, I considered that, spiders though they all were, they nevertheless differed from one another in their diverse shapes; so I sought their common character and found it. (quoted from Zimmermann 1981, 18 f.)

In this ordering activity, the author's restriction to the "most hideous among the insects," his discovery of the diversity within the class, and finally his identification of a general trait, all illustrate the "remarkable influence" it had on his "way of thinking."

Butterflies, worms, and caterpillars occupy a special place in children's hobby literature. The different developmental stages of these metamorphic creatures are easy to observe, hence their frequent featuring in lessons about Christian iconography (caterpillar, butterfly, resurrection). Moreover, any child can come across one. The illustration taken from a manual offering tips about starting such a collection (see Kaiserer 1802) displays an insect catcher's utensils and a cabinet with hundreds of little drawers for specimen storage (plates 43 and 44). The theologian and educator Georg Friedrich Seiler suggested in his "Elements for the Education of Future Schoolteachers, Preachers, Catechists, and Pedagogues" *(Grundsätze zur Bildung künftiger Volkslehrer, Prediger, Katecheten, und Pädagogen)* of 1783 that one should not only point out the worms, insects, flowers, and plants one happened upon during a walk, but have the children start "a little collection of them . . . and as soon as they can write, a little index" (quoted from Stach 1972, 27).

Visual demonstrations alone were not enough for stimulating learning; active, tactile dealings with the things of nature were also needed. Collecting laid a firm foundation for knowledge as well as for development of the mental faculties. It was on this basis—quite in the tradition of *Kunstkammer*—that comparative study could begin. Only from side-by-side placement of things, direct comparison of the objects themselves, might generality and regularity be found in the particular and thence an organizing overview obtained. A child could be schooled in analytical thought by "comparing many things of one kind or species, while viewing them side by side" (Stuve 1788, 290). It is only once his board was full that von Paula Schrank could notice the range of differing shapes of spiders. The child must reconcile facts by "comparison of extant notions" (Heusinger 1794, 74): "It is by their sensible connection [*leur rapport sensible*] . . . that he should appreciate all objects of Nature and all works of man" (Rousseau 1969, book 3, 294–95).

Thus the school cabinet or beginner's private collection occupied the wide scope of active accumulation, handling of objects, and grasping of nature.

Stoy also ascribed an importance to collecting that should not be underestimated. Although the Academy itself obviously could not properly accommodate natural specimens, the option of collating among the plates "suitable designs, pictures, or sketches" remained (1780–84, directions, 3). For this purpose and to enhance the aspect of the store, he thought "interleaving the copperplates with heavy sheets of paper" would be particularly commendable. So the available space these plates took up was expandible—even more so in the model of boxed cards, which inspires the user to continue acquiring material. The methodological trick lay in the interleaving of blank pages for whatever notes one wished to add. Nowhere was such continual expansion and collecting—even in the rigid linearity of book form—as widespread as in pedagogy. One "Universal Lexicon of Commercial Science" also suggested that, when a book was bound, "have writing paper interleaved and enter prices and new mercantile, statistical, and sundry notes" (Israel 1809, 1:xiv).

The aim of such education propagated via objects, data, and their consignment to compartments, drawers, tables, and boxes was described by Johann Beckmann in the foreword to his *Naturhistorie.*

> One must also help the pupils collect for themselves, however, whereby one provides them an occupation that is as pleasant as it is useful and with which one can ignite and fan their zeal. Although initially a collection only in name for a useful, at very least, harmless game, in adulthood, a love for collecting imbued in this way shall call them from the game table out into the field so as to extend their collection and marvel at the work of the Divine Creator. (1767)

Collecting is both basis and expression of a *Bürger*'s zeal, love of order, and industry.

The Faculties of the Reason and the Mind

With his Picture Academy Stoy wanted not merely to lay a foundation for knowledge but also, figuratively speaking, to erect a catalog inside a child's mind by which to organize collected scraps of knowledge. Without such a catalog children might "suddenly forget everything again, or retain nothing in proper order unless they pick out and remember the best while casting aside what is superfluous" (Stoy 1780–84, directions, 10). It was a matter of classification. Only a precise system in a child's mind might "identify all *naturalia,* even ones never before seen, and name them in a manner understandable to any naturalist" (Beckmann 1767, foreword). Collecting pictures was the groundwork upon which knowledge might be acquired, just as the sensational, haptic grasp at the material served primarily to transpose the box compartments into compartments of the mind. For, ultimately, the Picture Academy was a *Universalrepertorium* in which,

> since the beginning years of guidance onward, youth carries throughout its lifetime, as if in neatly ordered honeycomb cells, all the knowledge that reading, relations, profession, and experience has culled—whence, with little effort can be extracted again for use, all that has been learned and read, since the first year of instruction, of particular merit and remark. (Stoy 1780–84, directions, 8)

So Stoy's mental catalog even left room for individual experiences, social intercourse, and professional practice. "Many thousands of facts, truths, stories, and remarks gradually get lost again within our minds, because we do not in time build particular receptacles within which one can safely deposit and from which one can withdraw at will. Only *early imprinting* of these picture plates upon young minds can forestall this loss" (p. 8, original emphasis). But the teacher should respect the following criteria. For each tableau, nine blank pages should be bound into the volume—one per field. Each field had its own label related to its content, which Stoy referred to as a rubric: "Whatever the *élève* reads in the future that is worth the effort to remember and fits under these rubrics, this the teacher instructs him to enter there. . . . There may be nothing in the world that cannot be placed under these 468 rubrics, which can be still further subdivided" (p. 9). The tableaux effected such an *Impreßion* on the child's mind. At the same time, they existed materially and should be extended also by hand, in written form, until the pupil had outgrown the book. A youth would merely have to assimilate a "main catalog of all that has been heard and read" (p. 10); then—as in orderly housekeeping—items could not get lost anymore.

In Stoy's view, "deeper impressions" were formed in "combination with notions" (p. 9) inside the learning child's brain. This theory of lasting mental imprints stems from Descartes's theory of the memory: "As to the memory, I believe that that of material things depends on the vestiges that stay inside the brain after some image has been imprinted therein; and that that of intellectual things depends on some other vestiges that stay in the thought itself."[13] Descartes assigned great weight to these material engrams, which the mind reads off, so to speak. It was important that the structural changes inside the brain be organized in such a way that the stored images be recalled according to a set order. Material order in a child's environment was essential for material order and its representation in his or her brain.

An engraving was able to offer intense sensational experiences, and repeated viewing of it might stamp such impressions onto a child's brain. It was excellently suited to rendering a variety of impressions and promoting visualization. Stoy's Academy was one exceptional tool from this period, designed for making such mental impressions; other more general illustrated primers followed Basedow's conception. The historian, pedagogue, and later professor of politics August Ludwig von Schlözer, who evaluated Basedow quite critically, introduced in a quite different light the first German edition of the primer by the French magistrate Louis-René de Caradeuc de La Chalotais. In his foreword to that edition he praised its "order of succession of ideas such as are applied, expressed, and arrayed on a child's blank mind" (Schlözer 1771, 83). He outlined the plan that such an elementary work should follow: a progression

13. Descartes to P[ère Mesland] Leyden, 2 May 1644?, letter 347, in Descartes 1936–63, 4:114.

from the familiar to the unfamiliar; a classification of information in preparation for a systematization of reason. As the central storage place for learned material, the mind did not merely receive external impressions but also shaped them according to the "order of succession of ideas"; this was necessary for a child to remember what he had taken in.

This same idea is found in Stoy's Academy. A child's receptivity, the ordering of imprints on his or her mind, and mental tracks within the brain were related to the concept of "particular receptacles," which could be refilled arbitrarily—they needed only to be created. For this Stoy chose the tableau and its interlinked pictorial motifs. Once installed, this "meandering drawer" *(Windungsfach)* of the brain could continue to categorize and store additional insights. Similar vocabulary is employed in the organology of the anatomist and physician Franz Joseph Gall. He assigned specific abilities to defined regions of the human brain that purportedly could be read off the surface of the skull. Gall's theory was hotly debated at the time and, not surprisingly, was frequently charged with being unscientific. In retrospect one of his opponents, Jacob Fidelis Ackermann, was able to summarize the fascination of its suppositions:

> For these people [of the nonscientific class] it was entertaining to see all these human abilities and capacities suddenly thus embodied, and to know of certain little houses and drawers in which the mental faculties were all so nicely arranged. Accustomed to order, they were only too pleased to have [things work] in our brain cases as in their magazines of wares. (1806, 87)

With somewhat less metaphorical imagery, the philosopher and theologian Christian August Crusius described in his "Way toward Certainty and Reliability in Human Insight" *(Weg zur Gewißheit und Zuverläßigkeit der menschlichen Erkenntniß)* how the reason and the memory should be trained and which methods might be developed to coax the natural state into some order. The advantage it offered was that "what belongs to the meat of the matter ought not to be easily overlooked; throughout, one should encounter good order, the *imaginatio* should not be able to digress to foreign things or should easily be directed back onto the right path" (1747, 995). One possible linking element between these sketches—whether relating to brain cases, receptacles, or indices of the mind—is formalization followed by storage of experience, read or discovered fact. Mental tabletops and localities governed eighteenth-century notions of the human brain's receptive faculty and mental capacity.

Another recurrent motif accompanying the concepts of education, collecting, and even mental activity made use of the terms *sciagraphia* and *scrinium,* which likewise drew from the history of curiosity cabinets. Grimm defined *scrinium* as a shrine, a container, or a cabinet (1984, 15:1725) for storage of precious relics. For Major, on the other hand, it was a storage cabinet for a small-scale collection of *naturalia* (1674, chapter 5, § 7). Messerschmidt (1962–77, 3:216) also used the word to denote his travel boxes for classification and storage of his collected specimens. One hundred years earlier the naturalist and curator Johann Jakob Scheuchzer mentioned a *sciagraphia* that he intended to design, meaning the description, arrangement, and classification of the *Kunstkammer* in Zurich. Finally, Crusius later used this term as well: "Then one drafts of all these things a good sciagraph so that the imagination not stray" (1747, 1015), and this is exactly what, forty years later, Stoy proposed. Thus educators, cameralists, merchants, and physiologists alike employed concepts and metaphors from the history of collecting, with even such specialized terms as *sciagraphia* finding applications far removed from their time and place of origin.

Indexing as an intrinsic component of cabinets and boxes for the categorization of objects was hence supposed to be transposed to the mind. Upon translation into the two-dimensionality of a written page, the material referencing system of cabinets became superfluous for scientific inquiry. Consequently, educators took a dual course in their ordering of the world. Cabinets of natural specimens, machine models, and disparate visualizations of the waning eighteenth century educated a child's sensory faculties while simultaneously being functionalized for the mind and reason, for rationalized access to the world. Although the tradition of instructive collections continued to be cultivated in schools, universities, and even apothecary shops, in its ultimate form at the close of that century this tradition was nevertheless indentured to a rationally tabular analysis of the world.

Training for the World of the *Bürger*

A characteristic dichotomy developed, separating the private from the public spheres, which Habermas described as a "structural change of the public realm" (1962).[14] The attitude toward privacy and what is referred to as the public eye *(Öffentlichkeit)* changed not least in family life, along with its social embedding, as the structure of society was reshaped by the emergence

14. In this context I cannot go into the research on *Bürgertum,* which just recently has expanded considerably. One survey of current topics of research is provided by Roeck (1991); cf. Vierhaus 1981. Herrmann 1982 analyzes the complex of the *Bürger* and education; see also Dressen 1982, Grenz 1986, and Trefzer 1989.

of periodicals, for example (see Martens 1968), or the institutionalization of professional societies (see von Dülmen 1986). Simultaneously, we observe changes toward rational economy and property administration. Altogether, the essential factors in this process were early capitalistic commerce, trade, and the press, which latter turned information into a commodity.

Rationally conceived economic structures and state organizational forms refocused the family toward the private realm of society. This was accompanied by growing attention to the structure of private dwellings, as well as in education. In this ascent of *Bürgertum* four main groups emerged: merchants, the clergy, middle-class intellectuals—that is, holders of public office—and military officers. Proper education for their children included transmitting a consciousness appropriate to their social station: "Schools for raising scholars are there aplenty; schools too for raising soldiers: but no schools for the education of the gainful *Bürger:* by his multifarious industry, the upholder of the state." Thus Friedrich Resewitz opened his book "Raising the *Bürger* to Practice Common Sense and Industry for Common Benefit" (*Die Erziehung des Bürgers zum Gebrauch des gesunden Verstandes, und zur gemeinnützigen Geschäfftigkeit* 1773, 3). This pedagogue and theologian presented the way he envisioned proper preparation of children for full membership in middle-class society. His "School of Education" *(Schule der Erziehung)* set out to promote knowledge of the various midranking professions. A *Bürger* is here defined as an ambitious person regulated by a solid work ethic, reason, usefulness, individual merit, morality, and propriety (pp. 158 ff.). He is the "gainful *Bürger*" who ranks among those who nurture and implement, and upon whose "head and hands the whole of society rests" (p. 6). Evidently, this School of Education must convey more than what can be gleaned from reading assignments and class discussion. Training for the world of the *Bürger* also involved ingraining the work habit while developing manual strength and dexterity.

In this regard the activity of gathering and adding new items to a collection might relate to the commercial sense a child was supposed to acquire (alongside craft skills). Only by fostering assiduousness and practical thinking might the idea of a bourgeois stratum and its characteristic traits as a group be brought together with the idea of economic prosperity. "Diligence is a major element of our obligation toward the world. It is the driving force behind engaging the powers of invention, assessment, notice, inquiry, conservation, fabrication, improvement, and dissemination of what is useful to people, or behind our efforts to make ourselves and others adept at such purposes" (Basedow 1768, 36). Once developed, diligence almost automatically engendered all the other human faculties: assessment, conservation, and dissemination. They might need some correction, yet they laid the basis for a life of independence, decision making, and practical thinking. Proper education of children of this rank meant activating and encouraging the powers Basedow mentioned.

In this process, work and play could merge. Essentially, a child's desire should "be tuned toward acquisition and gain. Gain has great advantages over all other possessions. It is much more gratifying to the possessor because it has the attendant idea of accomplishment and is proof of individual ability and excellence" (Villaume 1785, 438). What Villaume here called "proof of individual ability" reflects the autonomous attitude of the *Bürger.* Self-sufficiency had to be realized not only in the private sphere but also at work, and "control over property" (Habermas 1962, 58) became one of the mainstays in the identity of the bourgeoisie. For the consensus was that every owner of property within a particular state also had an interest in its general welfare.[15]

The idea of individual autonomy and ownership lay at the heart of a child's collecting activity. Critical assessment and comparison were involved in this accumulation of objects into a single place, and untiring management and cultivation of acquired property. Children should "themselves be active, and exercise their bodily energy—let them sow, pot, and tend the plants, collect flowers, herbs, rocks, butterflies, etc. . . . To add stimulus and life from another aspect: interweave feelings of ownership into the business . . . , let [the child] start an herb collection, a tree nursery, or the like" (Stuve 1788, 268 f.). Johann Beckmann added that, rather than seek out the rarest and most precious plants, children should "look instead for the most common and native ones. . . . Whoever wishes to have a collection of natural samples, need only make a start with the native ones and be alert; there shall always be occasion to add to it" (1767, foreword).

Adults, as well as children, were encouraged to keep such collections. This independent manipulation of objects and their indexing, the collecting, naming, and identifying processes, together make an apt symbol for the self-sufficient ambition of the *Bürgertum.* Knowledge was no longer an exclusive privilege; it was free for all to discover for themselves in nature. The busy investigator—whether adult or child—had the opportunity to exploit the realm of natural history by his or her own observation and his or her own hand. For Johann

15. Toward the end of the eighteenth century neither a bourgeois class nor a bourgeois state as yet existed. The numerous petty and city states in the German-speaking region (see Bruford 1975, 11–19) did not permit such a structure. Even so, a nascent "*bürgerliche* society" was apparent (Wehler 1987, 236 f.), which made the heterogeneous strata combine, not least by the common educational and cultural means under discussion here (see Assmann 1993); the role of civil servants as state functionaries and members of this "*bürgerliche* society" should also be mentioned (see Bruford 1975, 222–54; Wehler 1987, 385 ff.).

Jakob Scheuchzer collecting was virtually obligatory. He criticized physicians in Switzerland "who care not a straw for botany," that is, those who neither collected, prepared, nor classified any plants.

> Instead of climbing the mountains found right in front of their noses, they scarcely look at them, let alone enthusiastically embrace this wonderful opportunity to botanize. In my view, they absolutely deserve to be thrown out of that herbal garden in which they are living, as it were, in deranged oblivion, in order to make room for other doctors more eager about research and the cultivation of botany. (quoted from Steiger 1927–30, 80)

Accordingly, whoever did not put the available capital to practical use was deranged. Thus the motivation for investigating nature was economic usefulness. Botanizing indigenous plants provided, besides private intellectual gain, the basis for economic exploitation of territory, as Linnaeus pointed out in 1753 (see Jahn and Senglaub 1978, 67).

Simple specimen collection is here directly related to commercial exploitation. In the same spirit that regards nature as capital just waiting to be appropriated, Stoy seized the opportunity to capitalize on the encyclopedic knowledge of the late eighteenth century. The production and sale of picture cards was intended to strengthen a child's accumulating potential, which was of direct relevance to the gradually emerging state citizen (in replacement of the civil servant; see Wehler 1987, 209). At the same time, Stoy was following his own educational guidelines. His career change from theologian to businessman (a change made by some of his contemporaries as well) mirrors the conquering of new spheres of activity and a changing professional approach. The training of the *Bürger* therefore cannot be regarded simply as a matter of child education, since it also figured crucially in the careers of many people toward the close of the eighteenth century.

This conceptual backdrop explains the pedagogical approach of encouraging pupils to collect and of employing the box as collection's essential material basis. This approach of habituating a child to order, conservation, and accumulation sprang from the notion that such external training simultaneously found its way into the recesses of the child's mind.

Tableau D

Plate 36. Frontispiece for *Angenehme Bilder-Lust* (1760)

Plate 37. Frontispiece for *Historischer Bilder-Saal* (Imhof 1697)

Plate 38. Boy standing before a picture wall (Stoy 1780–84, copperplate 7, field 9)

Plate 39. The *studiolo* of the Medicis (sixteenth century) (Liebenwein 1977, 122)

Plate 40. *Hofbibliothek* and *Kunstkammer* in Vienna
(detail, Valentini 1704, tableau 38)

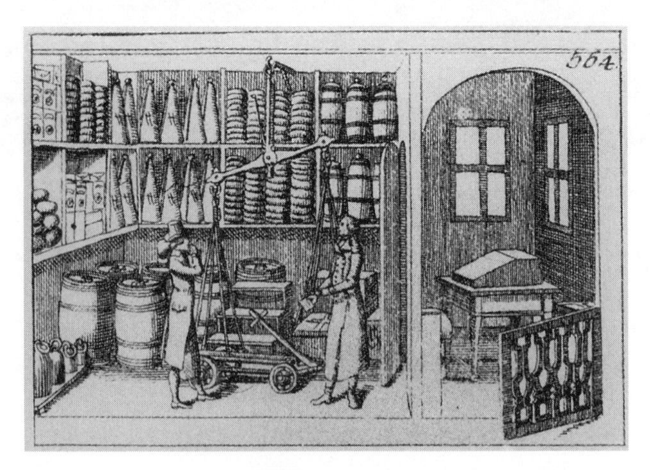

Plate 41. Shopkeeper's stockroom from a Bestelmeier catalog (1803)
(Junker and Stille 1984, 86)

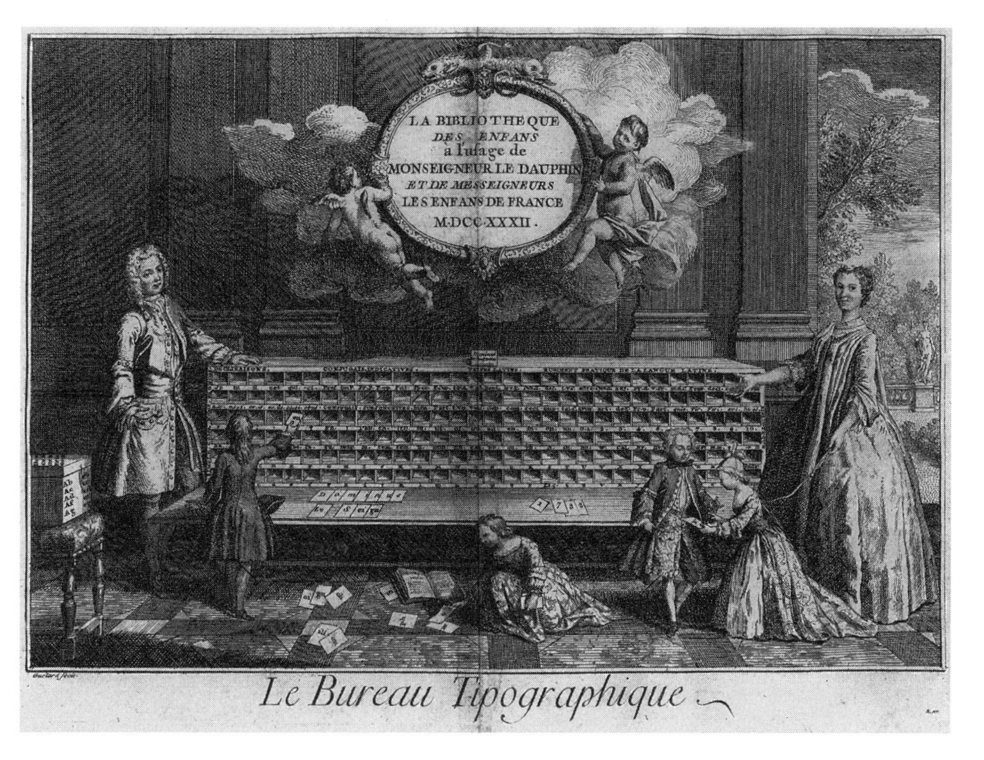

Plate 42. Frontispiece for *La Bibliothèque des enfans* (Dumas 1733)

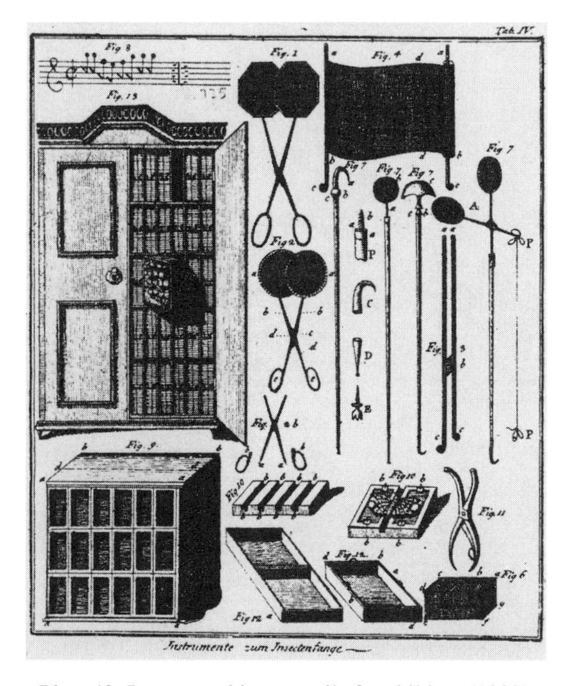

Plate 43. Insect-catching utensils for children (1802)
(Pressler 1980, 65)

Plate 44. A boy presenting his natural cabinet (1786)
(Alt 1965–66, 2:86)

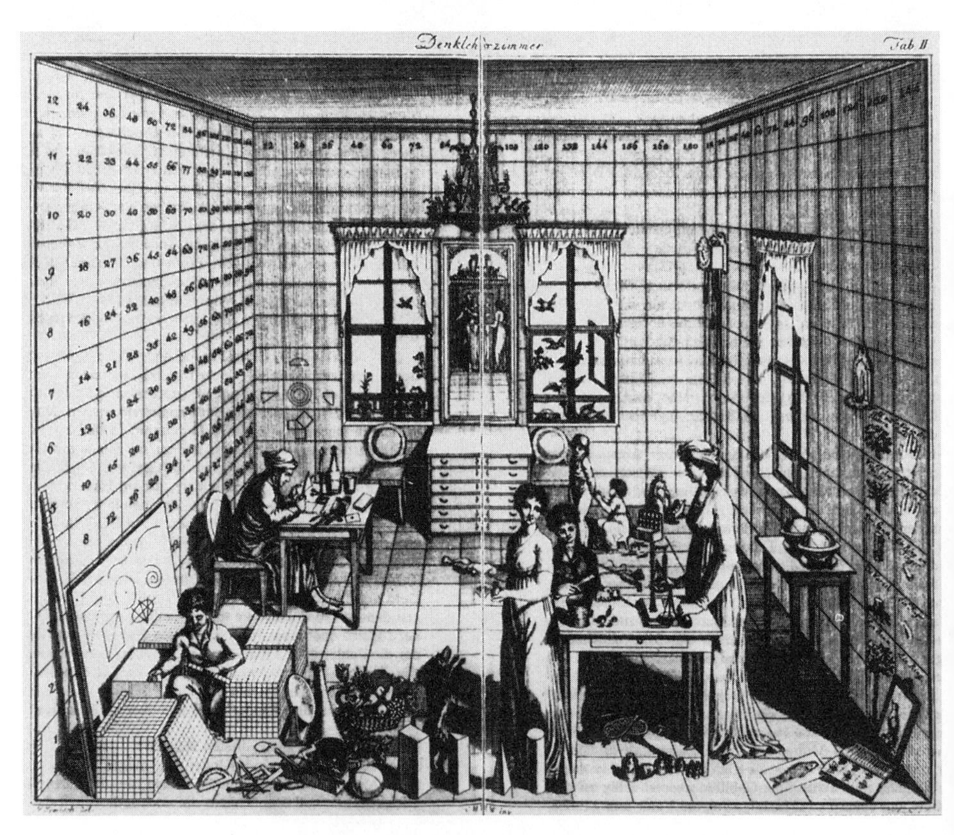

Plate 45. The “classroom of reason” *(Denklehrzimmer)* by Christian Heinrich Wolke
(1805) (Rutschky 1988, 154 f.)

CONCLUSION

The World Boxed In, Then and Now

Training for the world of the *Bürger* filled many aspects of a child's life during the eighteenth century. Children's literature, textbooks, toys, and edification tracts provide insight into the desired result. They were frequently written by men of multiple professions. Johann Siegmund Stoy, like Salzmann, Villaume, and Campe, abandoned the robe to become a pedagogical salesman. The problems he encountered with the church led him to seek his fortune in the new and expanding market for materials of instruction, without otherwise relinquishing the religious precepts. Means of instruction or "sensational aids" (*Versinnlichungsmittel,* Trapp 1787, 176) played a major part in this training. They culminated in the "classroom of reason" *(Denklehrzimmer)* of 1805 (plate 45) by the pedagogue Christian Heinrich Wolke.[1] The viewer looks into a perspectivally diminishing room in which a man, two women, and some children are sitting, standing, or playing. A miniature shopkeeper's counter can be seen in the back corner to the right; a few natural samples are lying about in the foreground along with some depictions of butterflies and a fish; a balance is set up on a table, and globes on a side table. All these recall Stoy's dedication copperplate and the realia of Philanthropinist instruction. In this classroom of reason, classifying and tabulating order reigns. More striking than these scattered objects is the impression of space given by the regular tiling covering the entire length of the walls. The boy at the front left-hand corner of the room is replicating it. This sectioning or cartography represents the elementary forms of calculation. It is a visualized interpretation of content already rationally converted into a spatial system of coordinates. By counting and measuring, the objects and people moving about the room may be localized and determined.[2] Drawn in relation to Stoy's Academy, the box compartments not only

1. Wolke was of the Philanthropinist school and was one of the collaborators on Basedow's *Elementarwerk;* the *Denklehrzimmer* and the goal of Wolke's didactic method, the "ultimate expression of the Philanthropinist methodology," is described by Stach (1970).

2. The individual family members demonstrate the various stages of abstract mathematical thought. Whereas the younger children busy themselves with the toy shop in the rear corner of the room, the older children are interested in the balance and are practicing spatial measurement; the father of the house is able to conduct his work in writing. This illustration retraces the course of rational development toward increasingly complex and abstract forms of reasoning.

correspond to the pictorial world stored inside it, but here also constitute the theoretical world, the worldview of a child by which he or she learns, grasps, and compares. The paterfamilia and his pupils have, so to speak, themselves climbed into the box.

This classroom of reason never materialized. Its furnishing, the way in which a group of adults and children study in harmony together in it, and its symbolical design indicate an ideal type. It is the epitome of the underlying concept of an educational enterprise like Stoy's. The Nuremberger's box of pictures joins the natural sample, a balance, blocks, and a tiny shop in its function as an aid to the senses. How and where these visualizing means of instruction could be enlisted into the service of educators is not always ascertainable. The criticism the Picture Academy received in contemporary reviews included, besides the weaknesses of the box medium and its pictorial content, the lack of control over eye and hand. Children could not concentrate properly, one stated in 1779, "their eyes light now on this picture, now that" (*Allgemeines Verzeichnis neuer Bücher* 1779, 4:613). Three years later another reviewer was more explicit: "The only concern is that fickle youth are wont to be unduly distracted by a sheet that depicts so much variety at once and let eyes and mind dart from field to field, without properly attending to any one of them" (*Allgemeine Bibliothek* 1783, 10:464). The available instructional tools would not effect consummate, carefully crafted fine-tuning of a child's faculties, nor were the demands of perceptual perfectionism so easily realized.

The concept of a box as a material and easily graspable equivalent of the complex world has not lost its attraction, however. The twentieth-century card box distributed by the Disney Company, a plastic catalog with movable partitions, is similarly designed for filing and storing picture cards: "cards for collecting knowledge." Its arrangement and even the wording of its instructions are astonishingly reminiscent of Stoy's box, without there being any direct link between them. Classified into fields like science, animals, professions, art, history, and space, each card carries a cartoon illustration and a photo on the front, and two columns of text on the back. This modern "flexible encyclopedia game" reiterates the Picture Academy's recommendation that it also "be used as an encyclopedia." "The included box allows your child to store the cards safely and to arrange them in the desired order"; the Academy likewise kept the learned material "in proper order." "The 'Disney Discovery Travels' collection is a fun source of knowledge for your child for years to come." Stoy also guaranteed the user of his Academy "a properly arranged and designed entertainment for many years to come." The Disney Company suggests, "Point out to your child that these cards contain many useful ideas and facts for school essays"; Stoy's cards provided material for short "exercises in reading, storytelling," essay writing, "questions, and drawings." The Disney Company closes its text with a reference to the separate shipments of cards: "As soon as you receive another card shipment, let your child file them away. . . . In that way he or she learns early on to distinguish between the separate topics."[3]

Even the language from 1993 agrees with Stoy's formulations dating back to the years 1779 through 1784. The box persists into the twentieth century as a fundamental means of educating a child in tackling the world's variety. The apparent paradox of the *Weltkasten,* a box for the world, that is, a surveyable, handy, and always accessible structured space for accommodating the complexity of our surroundings, thus persists, having acquired its own special stamp in the eighteenth century. The box, it seems, is an expression of that profound human need to organize the world around us, "to delimit, enclose, and label things" and "to make the abundance of actual fact manageable and comprehensible" (Friedell 1989, 59–60). Toward the end of the eighteenth century this coincided with the establishment of a social class that ranked order among the highest virtues. To his assertion "Order leads to all the virtues!" Lichtenberg appends, "but what, pray, leads to order?" (1973–75, 1:828). One response to this Göttingen scholar may be found in the material history of the Enlightenment, which remains alive today in the rank and file of things and their communion inside a box.

---

3. The quotes are from the "collector's instructions" provided with the first installment of the picture-card set Disney Entdeckungsreise from 1993; the sources of the Picture Academy quotes are Stoy 1780–84, directions, 13, 10; Stoy 1779, 30.

APPENDIX

Bibliographic Record of the Picture Academy

This record follows the guidelines laid down by Hans Ries in the 1982 issue of the journal *Die Schiefertafel* for the bibliographic treatment of illustration and visual representation in historical children's and adolescents' literature. Unless otherwise indicated, the description of the copperplate volume is based on the author's private copy; those of the text volumes are based on the copies in the Württembergische Landesbibliothek Stuttgart.

Bibliographic Data

Author: Johann Siegmund Stoy.

Title: *Bilder-Akademie für die Jugend.*

Subtitle: *Abbildung und Beschreibung der vornehmsten Gegenstände der iugendlichen Aufmerksamkeit—aus der biblischen und Profangeschichte, aus dem gemeinen Leben, dem Naturreiche und den Berufsgeschäften, aus der heidnischen Götter- und Alterthums-Lehre, aus den besten Sammlungen guter Fabeln und moralischer Erzählungen—nebst einem Auszuge aus Herrn Basedows Elementarwerke.*

Note on illustration and impression design: *In vier und funfzig Kupfertafeln und zweyen Bänden Erklärung herausgegeben von J. S. Stoy, Prof. der Pädagogik in Nürnberg.*

Edition: First.

Number of volumes: Two text volumes; one copperplate volume (initially published in nine separate issues).

Imprint: Nuremberg 1784, G. F. Six (printer).[1] The first two illustration installments were published by the Weigelsche Kunsthandlung in Nuremberg. From 1785 on it appeared in the author's publishing house (cf. the watermark).

Width: Volume 1: [various paginations] (6 unnumbered pages) dedication, 9–14 foreword [Vorrede], 1–16 directions [Anweisung], 1–8 sample [Probe], (4 unnumbered pages) notice [Nachricht] from 1782, I–IV connection between the images [Zusammenhang der Vorstellungen], 17–20 connection between the images, 9–12 connection between the images, (8 unnumbered pages) connection between the images, 1–580 text. Volume 2: (15 unnumbered pages) connection between the images, (1 unnumbered page) to the bookbinders [An die Herren Buchbinder], 581–1185 text, and 1186–1208 index.

Format: Text volumes: octavo (8°); copperplate volume: folio (4°), custom binding.

1. This "Raths- und Canzley-Buchdrucker" in Nuremberg printed the text volumes of the Picture Academy as well as the official malfeasance declaration *(Malefiz-Urthel)* of 1788.

Arrangement of the Front and Back Matter

Front matter: Text volumes: 1 blank sheet, no frontispiece, title page, dedication pages. Copperplate volume: Frontispiece, dedication page.

Typography

Type: German *(Luthersche Fraktur).*

Printing: Letterpress (relief).

Decorative elements: The opening plates of each tableau in the first and second volumes (e.g., "First Copperplate. The Creation of the World") have a block motif above the title at the running heads; a decorative band (garland) precedes all the other plates and closes each subject field.

Paper: Ribbed paper in the illustration volume. Watermark: "C W B" up to copperplate 36, or installment 6; the initials presumably stand for "Christoph Weigelsche Buchhandlung."

Illustrations

Number: 1 full-page dedication copperplate, 1 full-page frontispiece, 52 copperplate tableaux, each with nine fields (total 54 copperplates).

Design and shape: Full-page, broadside, uniform image outline with framing border, illustration separate from text, horizontal picture columns of nine bordered picture fields. Format: 270 max. × 200 max.

Reproduction technique: Intaglio printing and etching off copper; black/monochrome; individual copies in watercolor or aquarelle.

Technique of the original: probably pen and ink, copied from other illustrated works.

Artist signatures: Daniel N. Chodowiecki (1726–1801), Gottfried Chodowiecki (1728–81), Johann Georg Penzel (1754–1809), Johann Rudolf Schellenberg (1740–1806).

Engraver signatures: Johann Carl Bock (1757–1843), Carl Christian Glaßbach (1751–ca. 1789), F. Conrad Krüger, H. L. Möglich, Penzel (see above), Schellenberg (see above), Johann Friedrich Schleuen, and Jakob Samuel Walwert (1750–1815).

These signatures apply to the present copy; H. L. Möglich merely engraved the silhouettes of the six pedagogues for copperplate 41. With reference to the fourth installment Thanner indicates (1987, 294) that the engraving by Jakob Samuel Walwert was of such poor quality that it had to be redone by Schellenberg (see Stoy 1780–84, 13); the present copy includes the new copperplate. All engravers and artists, with the exception of Glaßbach, also worked on Basedow's *Elementarwerk.*

Structure and layout of the illustrations: The copperplates are composed of nine fields of various sizes, the central field being the largest, 101 max. × 135 max. The fields on either side of it, 101 max. × 69 max.; the fields in the upper and lower strips each 50 max. × 90 max.

Binding and Jacket

Sewing: Saddle stitching of the separate issues.

Dust jacket/cover: Custom binding of illustration volume with marble papered board and leather spine. Original binding: Separate fascicles (1–9) delivered in a simple blue blotting-paper-like cover with the legend, printed in black: "Bilder-Akademie für die Jugend. Ihro Kön. Hoheit dem Kronprinzen von Schweden zugeeignet. Achte Ausgabe nebst einigen Bogen Erklärung. Nürnberg 1783."[2] The bordering edges of these fascicle engravings are wider; they are otherwise reduced by trimming, binding, and so forth.

Price

Eight taler for commentary and plates; 2 taler for the text by itself (Stoy 1791, 5–6). Stoy already names this price in his notice of 1782, calculating 5 taler to 1 louis d'or; a separate installment by itself cost 1 taler. A distributing bookseller took 25% of this shelf price "with the explicit condition: not to sell the work for above this set price" (Stoy 1782, 4).

Retail outlets: The work was first made available in 1782 by the author at his "Bilder-Akademie-Contoir" in Henfenfeld. Orders could be placed with Möglich at Haymarket in Nuremberg, and it was offered for sale at "Post- und Intelligenz-Contoirs" or any art or book dealers in the city. Later the Picture Academy was marketed in Stoy's "Edukationshandlung."

Individual Description

The plates come in various bindings and can be grouped into three essential forms or uses:

1. Sewn and bound in hard cover, custom binding.
2. Complete copperplate tableaux mounted on cardboard backings, sometimes with decorative borders and hanging fixtures, and stored in cardboard sleeves.
3. Tableaux cut down into individual plates and mounted on appropriately sized cards that are filed in the specially designed box with separate partitions for each subject field.

In most cases, the physical condition of the copies is very good, disregarding brown spots from aging and soiled edges from use.

2. One such edition is part of the Strohbach Collection in Bielefeld.

Location of Originals

No information is available about the original manuscript. The text manuscript of the Academy was put up for auction by the Munich dealer Neumeister in 1983. Its current owner is not ascertainable.

Year of Publication

The illustrations appeared in nine installments of six copperplates each: installment 1, 1780 (1–6); installment 2, 1780 (7–12); installment 3, 1781 (13–18); installment 4, 1781 (19–24); installment 5, 1782 (25–30); installment 6, 1782 (31–36); installment 7, 1783 (37–42); installment 8, 1783 (43–48); installment 9, 1784 (49–52). The dedication plate and frontispiece were included with the ninth installment. According to Stoy (1782, 2), "a few explanatory sheets" were sent along with each delivery, which probably involved the sections entitled "connection between the images" that were subsequently bound at the front of the text volumes.

The text (vols. 1 and 2) appeared in 1784. The shorter textual supplements: Stoy's notice of 1782, his suggestion to the bookbinder, and the "connection between the images" are variously bound into the different copies.

Bound Copies

Staatsbibliothek zu Berlin, Department of Children's and Adolescents' Literature (call nos.: 19 ZZ 1307; B XIII, 66 R).

A. te Heesen, Berlin.

Strohbach Collection, Bielefeld.

Bibliothek des Instituts für Jugendbuchforschung, Frankfurt (call nos.: Sq 5/32, S5 448).

Universitäts- und Landesbibliothek Sachsen-Anhalt, Halle (call no.: Gc 44).

T. Brüggemann Collection, Cologne.

Bayerische Staatsbibliothek, Munich (call no.: Paed. Pr. 320a).

Bibliothek des Germanischen Nationalmuseums, Nuremberg (call no.: W 16502/20b).

Stadtbibliothek, Nuremberg (call nos.: Amb2103, 1 and 2), Manuscripts Department (call no.: Nor. K171).

Österreichische National Bibliothek, Vienna (call nos.: 65.H.50, 65.J.6).

Württembergische Landesbibliothek, Stuttgart (call nos.: MC HBFa 1218, MC HBF 4657-1, 2).

Sammlung historischer Kinder- und Jugendbücher Hubert Göbels und Othmar Hicking.

Box Versions

J. Garber, Halle.
Bibliothek des Germanischen Nationalmuseums, Nuremberg (call no.: W 16502/20a).
R. Prinz zur Lippe, Oldenburg.

Reprinted Editions

1789 *Nouveau manuel élémentaire ou explication d'une suite d'estampes pour l'instruction de la jeunesse trad. de l'Allemand par Perrault.* Nuremberg: Author-publisher; Winterthur: H. Steiner (2d ed. 1790; 3d ed. 1814).
1793 *Kleine Bilderakademie für leselustige und lernbegierige Söhne und Töchter.* Berlin: E. Felisch.
1814 *Elementarisches Handbuch, zum Unterricht der Jugend oder die ersten Begriffe in allen Arten dargestellt.* Breslau: W. G. Korn.
1999 *Bilder-Akademie für die Jugend.* 3 vols., edited by Reinhard Stach and Othmar Hicking, with commentary by Othmar Hicking. Nachdrucke historischer Kinder- und Jugendbücher. Andernach: KARI-Verlag.
The 1999 edition is the first comprehensive reprinting of the Picture Academy. This list of reprintings and new editions is probably not comprehensive. More detailed information about the above mentioned different editions is provided by Thanner (1987, 356 ff.).

Contemporary Reviews

Allgemeine Bibliothek 1780, 8:383–90; 1783, 10:463–68.
Allgemeine Deutsche Bibliothek 1782, vol. 49, no. 1, 221–24.
Allgemeines Verzeichnis neuer Bücher 1779, 4:612–14; 1782, 7:380 f.; 1783, 8:425 f.; 684 f.; 834 f.
Baur 1790, 483.
Bertuch 1790, 2.
Neue Bibliothek der schönen Wissenschaften und der freyen Künste 1780, 25:144 f.
Nürnbergische gelehrte Zeitung, 28 May 1779, 103 f.; 22 October 1779, 693–95; 13 October 1780, 662–64; 12 February 1782, 99 f.; 13 March 1781, 165 f.; 15 November 1782, 735 f.; 4 July 1783, 425; 17 September 1784, 603 f.

Modern Literature about the Author and the Work

For the sake of completeness, these include cursory comments and bibliographic classifications; mere mentions of Stoy and his work in exhibition catalogs or literature surveys are omitted.

Bravo-Villasente 1977, 42.
Brüggemann 1979, 472–75.
Brüggemann and Ewers 1982–91, 1099–114.
Dierks 1965, 30.
Doderer and Müller 1973, 13, 45.
Dyhrenfurth-Graebsch 1951, 37.
Fricke 1886, 137.
Göbels 1965, 42–45; 1980, 124–27.
Göhring 1904, 65.
Hauswedell and Voigt 1977, 211, 224.
Hobrecker 1924, 29 f., 137.
Hruby 1986, 170.
Köberle 1972, 172, 192.
Köster 1972, 17.
Krebs 1929.
Pilz 1967, 335–39.
Rammensee 1961, 181.
Reuter 1994.
Richter 1987, 84, 99.
Ringshausen 1976, 461–64.
Schug 1988, 151, 331 f.
Strobach 1978, 35, 38.
Thanner 1987, 291–368.
Uphaus-Wehmeier 1984, 186.
Wegehaupt 1979, 238 f.; 1992, 66, 69.
Wild 1990, 90, 93.

BIBLIOGRAPHY

Unpublished Sources

Pfarrarchiv Henfenfeld
- 156: Catalog of Books of the Parish Library 1556–1869
- VIII/11: Description of Henfenfeld Parish for 1914
- VIII/70: Files of the Royal Bavarian Protestant Pastorate of Henfenfeld, the Parish Chronicle and Parish Description
- VIII/72: Parish Register of the Lutheran Parish of Henfenfeld for 1864
- XVI/2b: Church Tower Repair 1780

Staatsarchiv Nürnberg
- Rep. 52a, Reichsstadt Nürnberg, Manuscripts 222: Wappenbuch: Coat of arms of the Stoy family
- Rep. 60a, Reichsstadt Nürnberg, Verlässe des Innern Rats, no. 3975: 17 May 1771 (re dismissal of Preacher Stoy)
- Rep. 60a, Reichsstadt Nürnberg, Verlässe des Innern Rats, no. 4125: Saturday, 23 November 1782 (re resignation of pastorship)
- Rep. 60d, Reichsstadt Nürnberg, Verlässe der Herren Älteren, no. 70: 12 March 1771 (re lapses by Preacher Stoy)
- Rep. 92, Reichsstadt Nürnberg, Testament, no. 1118: Testament of J. S. Stoy and B. M. Stoy

Staatsbibliothek zu Berlin Preußischer Kulturbesitz (Manuscripts Department)
- Darmstaedter Collection, 2d1778(6): 6 letters from J. S. Stoy to D. N. Chodowiecki

Stadtarchiv Nürnberg
- A6 Mandat 17 September 1771: Mandate on the general rise of prices
- City Archive Papers, Bz/E5b: Escutcheon of J. S. Stoy by M. Tyroff

Stadtbibliothek Nürnberg (Manuscripts Department)
- Will Papers, III, 457 Autogr.: Letter from J. S. Stoy to G. A. Will

Zentralbibliothek Zürich (Manuscripts Department)
- Autograph Collection OH: Letter from J. S. Stoy to unknown addressee

Primary Sources

Ackermann, Jacob Fidelis. 1806. *Die Gall'sche Hirn-, Schedel- und Organenlehre vom Gesichtspunkte der Erfahrung aus beurtheilt und widerlegt.* Heidelberg and Frankfurt am Main.

Adelung, Johann Christoph. 1771. *Unterweisung in den vornehmsten Künsten und Wissenschaften, zum Nutzen der niedern Schulen.* Frankfurt am Main and Leipzig.

Allgemeine Bibliothek für das Schul- und Erziehungswesen in Deutschland. 1773–86. Nördlingen.

Allgemeine Deutsche Bibliothek. 1765–96. Berlin and Stettin.

Allgemeines Verzeichnis neuer Bücher mit kurzen Anmerkungen. Nebst einem gelehrten Anzeiger. 1776–84. Leipzig.

Angenehme Bilder-Lust: Der lieben Jugend zur Ergötzung also eingerichtet. 1760. Nuremberg.

Basedow, Johann Bernhard. 1768. *Die ganze natürliche Weisheit im Privatstande der gesitteten Bürger.* Altona and Halle.

———. 1770. *Das Methodenbuch für Väter und Mütter der Familien und Völker.* Altona and Bremen.

———. 1771. *Agathokrator, oder Von der Erziehung künftiger Regenten.* Leipzig.

———. [1785] 1909. *Elementarwerk.* Edited by Th. Fritzsch. Reprint of 2d ed. with copperplates by Chodowiecki et al. 3 vols. Leipzig.

Baur, Samuel. 1790. *Charakteristik der Erziehungsschriftsteller Deutschlands. Ein Handbuch für Erzieher.* Leipzig.

Beckmann, Johann. 1767. *Anfangsgründe der Naturhistorie.* Göttingen and Bremen.

———. 1911. *Schwedische Reise nach dem Tagebuch der Jahre 1765–1766.* Edited by Th. M. Fries. Uppsala.

Bertuch, Friedrich Justin. 1790. Plan, Ankündigung, and Vorbericht des Werks in *Bilderbuch für Kinder.* Weimar and Gotha.

———. 1792–1813. *Bilderbuch für Kinder enthaltend eine angenehme Sammlung von Thieren, Pflanzen, Blumen, Früchten, Mineralien, Trachten und allerhand andern unterrichtenden Gegenständen aus dem Reich der Natur, der Künste und Wissenschaften.* 8 vols. Weimar and Gotha.

Beschäftigungen für die Jugend aller Stände zur Gewöhnung an zweckmäßige Thätigkeit, zur erheiternden Unterhaltung so wie zur Anregung des Kunst- und Gewerbesinnes. 1834–40. Stuttgart.

Bestelmeier, Georg Hieronimus. 1793–96. *Pädagogisches Magazin zur lehrreichen und angenehmen Unterhaltung für die Jugend.* Nuremberg.

Blasche, Bernhard Heinrich. 1797. *Der Papparbeiter, oder Anleitung in Pappe zu arbeiten. Vorzüglich Erziehern gewidmet.* Schnepfenthal.

Büsching, Anton Friedrich. 1775. *Unterricht in der Naturgeschichte, für diejenigen, welche noch wenig oder gar nichts von derselben wissen.* Berlin.

Busse, Friedrich Gottlieb von. 1795–96. *Kenntnisse und Betrachtungen des neuern Münzwesens, für Deutschland.* 2 vols. Leipzig.

Campe, Joachim Heinrich. 1778. *Sammlung einiger Erziehungsschriften.* 2 vols. Leipzig.

Chodowiecki, Daniel, and G. C. Lichtenberg. 1977. *Der Fortgang der Tugend und des Lasters: Daniel Chodowieckis Monatskupfer zum Göttinger Taschenkalender mit*

Erklärungen Georg Christoph Lichtenbergs, 1778–1783. Edited by I. Sommer. 2d ed. Frankfurt am Main.

Cicero, Marcus Tullius. 1988. *De oratore.* Translated by E. W. Sutton and edited by H. Rackham. 2 vols. Cambridge, Mass.

Comenius, Johann Amos. 1685. *Orbis sensualium pictus quadrilinguis.* Reprint of the original edition. Prague, 1979.

Crusius, Christian August. 1747. *Weg zur Gewißheit und Zuverläßigkeit der menschlichen Erkenntniß.* Leipzig.

Descartes, René. 1936–63. *Correspondance.* Vol. 4 of *Oeuvres de Descartes,* edited by Charles Adam and Gérard Milhaud. Paris.

———. [1637] 1964–74. *Discours de la méthode pour bien conduire sa raison et chercher la vérité dans les Sciences. Plus la Dioptrique, les Météores et la Géométrie. Qui sont des essais de cette méthode.* Leyden. Vol. 6 of *Oeuvres de Descartes,* edited by Charles Adam and Paul Tannery. Paris.

Dumas, Louis. 1733. *La Bibliothèque des enfans, ou Les premiers élémens des Lettres, content le sistème du Bureau tipographique.* Paris.

Ebert, Johann Jacob. 1773. *Natürliche Geschichte, aus seiner näheren Unterweisung in den philosophischen und mathematischen Wissenschaften.* Karlsruhe.

———. 1775. *Kurze Unterweisung in den Anfangsgründen der Naturlehre zum Gebrauch der Schulen.* Leipzig.

———. 1776–78. *Naturlehre für die Jugend.* 3 vols. Leipzig.

Engel, Johann Jakob. 1785–86. *Ideen zu einer Mimik.* 2 vols. Berlin.

Ernesti, Johann Heinrich Martin. 1778. *Praktische Unterweisung in den Wissenschaften für die kleine Jugend durch Muster, meisten moralischen Inhalts.* Nuremberg.

Evenius, Sigmund. 1636. *Christliche Gottselige Bilder Schule: Das ist, Anführung der ersten Jugend zur Gottseligkeit, in und durch Biblische Bilder, auß und nach den Historien, Sprüchen der Schrifft, Einstimmung deß Catechismi, und nützlichen Gebrauch erkläret.* Jena.

Goethe, Johann Wolfgang von. [1800] 1895. Die guten Weiber. In *Goethes Werke. Hrsg. im Auftrag der Grossherzogin Sophie von Sachsen,* pt. 1, vol. 18, 275–312. Weimar.

Grimm, Jacob, and Wilhelm Grimm. 1984. *Deutsches Wörterbuch.* 33 vols. Munich.

Gründlich- und Nützlicher Unterricht von der Haußhaltungs-Kunst und guten Haußhaltern. 1720. Leipzig and Halle.

Gütle, Johann Conrad. 1790. *Beschreibung verschiedener Elektrisirmaschienen zum Gebrauch für Schulen.* Leipzig and Nuremberg.

———. 1791. *Versuche, Unterhaltungen und Belustigungen aus der natürlichen Magie zur Lehre, zum Nutzen und zum Vergnügen bestimmt.* Leipzig and Jena.

Gutsmuths, Johann Christoph Friedrich. 1785. *Zusammenkünfte am Atlas zur Kenntniß der Länder, Völker und ihrer Sitten herausgegeben für die Jugend.* Gotha.

Halle, Johann Samuel. 1761–79. *Werkstäte der heutigen Künste, oder Die neue Kunsthistorie.* 6 vols. Brandenburg and Leipzig.

——. 1779–80. *Kleine Encyclopedie, oder Lehrbuch aller Elementarkenntnisse, worinnen die Hauptbegriffe von allen Wissenschaften, von allen nützlichen Künsten und von allen Dingen gegeben werden, die auf die bürgerliche Gesellschaft einen Einfluß haben.* Berlin and Leipzig.

Happel, Eberhard Werner. [1683–91] 1990. *Größte Denkwürdigkeiten der Welt oder Sogenannte Relationes Curiosae.* Reprint edited by U. Hübner and J. Westphal. Berlin.

Heusinger, Johann Heinrich Gottlieb. 1794. *Beytrag zur Berichtigung einiger Begriffe über Erziehung und Erziehungskunst.* Halle.

Hirschfeld, Christian Cay Lorenz. 1775. *Von der moralischen Einwürkung der bildenen Künste.* Leipzig.

Hübner, Johann. 1759. *Zweymal zwey und fünfzig auserlesene Biblische Historien aus dem Alten und Neuen Testamente, der Jugend zum Besten abgefasset.* Zion.

Hufeland, Christoph Wilhelm. 1812. *Geschichte der Gesundheit nebst einer physischen Karakteristik des jetzigen Zeitalters. Eine Vorlesung in der Königl. Akademie der Wissenschaften zu Berlin.* Berlin.

Imhof, Andreas Lazarus von. 1697. *Neu-eröffneter historischer Bilder-Saal.* Vol. 1. Nuremberg.

Israel, Moses, et al. 1809. *Universal-Lexicon der Handlungswissenschaften.* Vol. 1. Leipzig.

Justi, Johann Heinrich Gottlob von, ed. 1762–1805. *Schauplatz der Künste und Handwerke, oder Vollständige Beschreibung derselben, verfertiget oder gebilliget von denen Herren der Academie der Wissenschaften zu Paris.* 21 vols. Berlin.

Kaempf, Johann. 1784. *Abhandlung von einer neuen Methode, die hartnäckigsten Krankheiten, die ihren Sitz im Unterleibe haben, besonders die Hypochondrie, sicher und gründlich zu heilen.* Dessau and Leipzig.

Kaiserer, Jacob. 1802. *Beschäftigungen für die Jugend in ihren Erholungsstunden. Ein Handbuch für Aeltern und Erzieher; worin Anleitung gegeben wird, wie man Säugethiere, Vögel, Amphibien, Fische, Insecten und Würmer fangen, Pflanzen einsammeln, und diese sowohl als jene für ein Naturalienkabinett aufbewahren könne; ferner wie man Münzen oder Medaillen in verschiedenen Materien abdrucken oder abgießen könne; u. dgl. mehr.* Vienna.

Der Kinderfreund. Ein Wochenblatt. 1780–82. 12 vols. 3d impression. Leipzig.

Kleines Arbeitsbuch für Kinder in den Erholungsstunden, oder angenehme und nützliche Beschäftigungen für die Jugend die wenig Kostenaufwand verursachen, und zur körperlichen und moralischen Bildung sehr wirksam sind. 1810. Pirna.

Kohfeldt, Gustav, and W. Ahrens, eds. 1919. *Ein Rostocker Studenten-Stammbuch von 1736/37: Zur Feier des 500jährigen Bestehens der Universität Rostock herausgegeben.* Rostock.

Köhler, Johann David. 1762. *Anweisung für Reisende Gelehrte: Bibliothecken, Münz-Cabinette, Antiquitäten-Zimmer, Bilder-Säle, Naturalien- und Kunst-Kammern u.d. m. mit Nutzen zu besehen.* Frankfurt am Main and Leipzig.

Lavater, Johann Caspar. 1775–78. *Physiognomische Fragmente zur Beförderung der Menschenkenntnis und Menschenliebe.* 4 vols. Leipzig and Winterthur.

——. 1794. *Regeln für Kinder, durch Beyspiele erläutert von Johann Michael Armbruster. Zum Gebrauch in Schulen und beym Privatunterricht.* St. Gallen.

Leipziger Wochenblatt für Kinder. 1773–83. Frankfurt an der Oder and Leipzig.

Lichtenberg, Georg Christoph. 1973–75. *Schriften und Briefe.* Vols. 1 and 2, *Sudelbücher.* Edited by W. Promies. 2d ed. Munich.

Linck, Johann Heinrich. 1783–87. *Index Musaei Linckiani, oder Kurzes systematisches Verzeichniß der vornehmsten Stücke der Linckischen Naturaliensammlung zu Leipzig.* 3 vols. Leipzig.

Ludovici, Carl Günther. 1752–56. *Eröffnete Akademie der Kaufleute, oder Vollständiges Kaufmanns-Lexicon, woraus sämtliche Handlungen und Gewerbe, mit allen ihren Vortheilen, und der Art, sie zu treiben, erlernet werden können.* 5 vols. Leipzig.

Major, Johann Daniel. 1674. *Unvorgreiffliches Bedencken von Kunst- und Naturalien-Kammern insgemein.* Kiel.

Malefiz-Urthel über Johann Philipp Feigel. 1788. Nuremberg.

Marperger, Paul Jacob. 1708. *Das in Natur- und Kunst-Sachen Neu-eröffnete Kauffmanns-Magazin.* Hamburg.

Marx, Johann Jacob. 1709. *Neu- viel- vermehrte und verbesserte Teutsche Material-Kammer.* Nuremberg.

Merian, Matthäus. 1630. *Biblia, das ist die gantze Heilige Schrifft durch D. Martin Luther verteutscht.* Strasbourg.

——. N.d. *Bybel Printen, vertoonende de voornaemste Historien der Haylige Schrifture konstigh afgeteelt.* Amsterdam.

Messerschmidt, Daniel Gottlieb. 1962–77. *Forschungsreise durch Sibirien, 1720–1727.* Edited by E. Winter et al. 5 vols. Quellen und Studien zur Geschichte Osteuropas, no. 8. Berlin.

Messkatalog. 1779–1784. *Allgemeines Verzeichnis derer Bücher, welche in der Frankfurter und Leipziger Ostermesse [od. Michaelismesse] des 1779 [–1784] Jahres entweder ganz neu gedruckt, oder sonst verbessert, wieder aufgelegt worden sind, auch ins künftige noch herauskommen sollen.* Leipzig.

Neickelius, C. F. [Caspar Friedrich Jencquel]. 1727. *Museographia oder Anleitung zum rechten Begriff und nützlicher Anlegung der Museorum oder Raritäten-Kammern.* Leipzig and Breslau.

Neue Bibliothek der schönen Wissenschaften und der freyen Künste. 1765–1806. Leipzig.

Nicolai, Christoph Friedrich. 1783. *Beschreibung einer Reise durch Deutschland und die Schweiz, im Jahre 1781. Nebst Bemerkungen über Gelehrsamkeit, Industrie, Religion und Sitten.* Vols. 1 and 2. Berlin and Stettin.

Nürnbergische gelehrte Zeitung. 1777–1789. Nuremberg.

Pluche, Noël Antoine. 1733–39. *Le Spectacle de la nature: Entretiens sur les particularités de l'histoire naturelle. Qui ont paru les plus propres à rendre les Jeunes Gens curieux, & à leur former l'esprit.* Utrecht.

——. 1760–66. *Schau-Platz der Natur, oder Unterredungen von der Beschaffenheit und den Absichten der natürlichen Dinge, wodurch die Leser zu weiterm Nachforschen aufgemuntert, und auf richtige Begriffe von der Allmacht und Weißheit Gottes geführet werden.* 8 vols. Frankfurt am Main and Leipzig.

Raff, Georg Christian. 1778. *Naturgeschichte für Kinder zum Gebrauch in Stadt- und Landschulen.* Göttingen.

Resewitz, Friedrich Gabriel. 1773. *Die Erziehung des Bürgers zum Gebrauch des gesunden Verstandes, und zur gemeinnützigen Geschäfftigkeit.* Copenhagen.

Rohr, Julius Bernhard von. 1718. *Einleitung zur Staatsklugheit, oder Vorstellung wie Christliche und weise Regenten zur Beförderung ihrer eigenen und ihres Landes Glückseeligkeit ihre Unterthanen zu beherrschen pflegen.* Leipzig.

Roth, Johann Ferdinand. 1800–1802. *Geschichte des Nürnbergischen Handels. Ein Versuch.* 4 vols. Leipzig.

Rousseau, Jean-Jacques. [1762] 1969. *Émile, ou De l'éducation.* Reprint edited by Charles Wirtz and Pierre Burgelin. Paris.

Rudolph, Daniel Gottlob. 1766. *Hand-Buch oder kurze Anweisung wie man Naturalien-Sammlungen mit Nutzen betrachten soll.* Leipzig.

Salzmann, Christian Gotthilf. 1793. *Reisen der Salzmannischen Zöglinge.* Vol. 6. Leipzig.

Sandrart, Johann Jacob. N.d. *P. Ovidii Nasonis Metamorphosis libri VII.* Nuremberg.

Schellenberg, Johann Rudolf. 1779. *60 Biblische Geschichten des Neuen Testament.* 2 vols. Winterthur.

Schlözer, August Ludwig von. 1771. Foreword of the German editor, in *Versuch über den Kinder-Unterricht von L. Renatus de Caradeuc de la Chalotais,* iii–xciii. Göttingen and Gotha.

——. 1779–1806. *Vorbereitung zur Weltgeschichte für Kinder.* 2 vols. Göttingen.

Schröckh, Johann Matthias. 1786–97. *Allgemeine Weltgeschichte für Kinder.* 4 vols. 2d ed. Leipzig.

Seiferheld, Georg Heinrich. 1787–99. *Sammlung electrischer Spielwerke für junge Electriker.* 8 installments. Nuremberg and Altdorf.

Sprengel, Peter Nathan. 1767–95. *Handwerke in Tabellen.* 17 vols., continued by O. L. Hartwig. Berlin.

Stoy, Johann Siegmund. 1778–81. *Der goldene Spiegel. Ein Lesebuch für Kinder.* 3 vols. Nuremberg.

——. 1779. *Ausführlicher Entwurf des Buches das unter dem Titel: Neue Bilder-Academie für die Jugend von den berühmtesten Künstlern Deutschlands verfertiget und von der Weigelischen Kunsthandlung in Nürnberg verlegt wird.* Nuremberg.

——. 1780–84. *Bilder-Akademie für die Jugend.* 3 vols. Nuremberg.

——. 1782. Nachricht. In *Bilder-Akademie für die Jugend.* Nuremberg.

——. 1788. *Kleine Biographie für die Jugend. Aus dem pädagogischen Kabinette des Professors Stoy in Nürnberg.* Nuremberg.

——. 1789. *Der achtzehnte Merz. Ein Wort zu seiner Zeit an die gesamte Bürger-*

schaft, des heute vor einem Jahre aufs Rad geflochtenen Mörders und Totengräbers Feigel. Nuremberg.

———. 1791. *Weitläuftige Beschreibung des pädagogischen Kabinets, welches ich zur Erleichterung der Erziehung und zur lehrreichen Beschäftigung und Belustigung der Jugend angelegt habe.* Nuremberg.

———. 1801. *Kurzer Entwurf einer ganz unerhörten Geschichte, allen rechtschaffenen und zum Wohlthun geneigten lieben Mitbürgern zur Beherzigung und Warnung vorgestellt.* Nuremberg.

Stuve, Johann. 1788. Über die Nothwendigkeit, Kinder zu anschauender und lebendiger Erkenntniß zu verhelfen, und über die Art, wie man dies anzufangen habe. *Allgemeine Revision des gesammten Schul- und Erziehungswesens von einer Gesellschaft praktischer Erzieher* 10:163–444.

Sulzer, Johann Georg. 1771–74. *Allgemeine Theorie der Schönen Künste in einzeln, nach alphabetischer Ordnung der Kunstwörter auf einander folgenden, Artikeln abgehandelt.* 2 vols. Leipzig.

Tiedemann, Dietrich. 1787. Beobachtungen über die Entwickelung der Seelenfähigkeiten bei Kindern. *Hessische Beiträge zur Gelehrsamkeit und Kunst* 2:313–33, 486–502.

Trapp, Ernst Christian. 1787. Vom Unterricht überhaupt. Zweck und Gegenstände desselben für verschiedene Stände. Ob und wie fern man ihn zu erleichtern und angenehm zu machen suchen dürfe? Allgemeine Methoden und Grundsätze. *Allgemeine Revision des gesammten Schul- und Erziehungswesens von einer Gesellschaft praktischer Erzieher* 8:1–210.

Trembley, Abraham. 1775. *Instructions d'un père à ses enfans, sur la nature et sur la religion.* 2 vols. Geneva.

Tyroff, Martin. 1783. *Sammlung von Wappen verschiedener Staende ausser dem Adel.* Nuremberg.

Valentini, Michael Bernhard. 1704. *Museum Museorum, oder Vollständige Schau-Bühne aller Materialien und Specereyen nebst deren natürlichen Beschreibung, Election, Nutzen und Gebrauch.* 3 vols. Frankfurt am Main.

Vieth, Gerhard Ulrich Anton. 1798–1809. *Physikalischer Kinderfreund.* 10 vols. Leipzig.

Villaume, Peter. 1785. Allgemeine Theorie, wie gute Triebe und Fertigkeiten durch die Erziehung erwekt, gestärkt und gelenkt werden müssen. *Allgemeine Revision des gesammten Schul- und Erziehungswesens von einer Gesellschaft praktischer Erzieher* 4:1–604.

———. 1787. Von der Bildung des Körpers in Rücksicht auf die Vollkommenheit und Glückseligkeit des Menschen, oder über die physische Erziehung insonderheit. *Allgemeine Revision des gesammten Schul- und Erziehungswesens von einer Gesellschaft praktischer Erzieher* 8:211–492.

Voit, Johann Peter. 1774–83. *Schauplatz der Natur und Künste, in vier Sprachen, deutsch, lateinisch, französisch, und italienisch.* 10 vols. Vienna.

Von hiesigen Professoren, und besonders am Egydianischen Auditorium. 1787. *Beyträge zur Geschichte der Stadt Nürnberg* 2:233–64.

Waldau, Georg Ernst. 1779. *Verzeichnisse und Lebensbeschreibungen aller Herrn Geistlichen in der Reichsstadt Nürnberg von 1756 bis zum Schluß des Jahres 1779 fortgesetzt.* Nuremberg.

Weigel, Christoph. 1695. *Biblia Ectypa. Bildnußen auß Heiliger Schrift des Alt- und Neuen Testaments, in welches Geschichte und Erscheinungen deutlich und schriftmäßig zu Gottes Ehre und Andächtiger Seelenerbaulicher Beschauung vorgestellet werden.* Augsburg.

——. 1700. *Die Welt in einer Nuß, oder Die Historien vom Anfang der Welt samt deren Zeit-Rechnung biß auff unsere Zeit auf eine besondere und ganz leichte Art kurz zusammen zufassen oder ausgebreitet in einem Augenblick auf einer einigen Tafel zu wiederhohlen fürgeschrieben und fürgebildet.* Nuremberg.

——. 1708. *Historiae celebriores veteris testamenti iconibus repraesentatae et Ad excitandas bonas meditationes selectis Epigrammatibus exornatae in lucem datae.* Nuremberg.

Will, Georg Andreas. 1766. *Die Grösse und Mannigfaltigkeit in den Reichen der Natur und Sitten nach der Absicht des Schöpfers von je her verbunden.* Nuremberg.

Will, Georg Andreas, and C. C. Nopitsch, eds. 1808. *Nürnbergisches Gelehrten-Lexikon, oder Beschreibung aller Nürnbergischen Gelehrten beyderley Geschlechts . . . in alphabetischer Ordnung verfasset.* Suppl. 4. Nuremberg and Altdorf.

Winckelmann, Johann Joachim. [1755] 1995. Gedanken über die Nachahmung der griechischen Wercke in der Mahlerey und Bildhauer-Kunst. Reprint in *Frühklassizismus. Position und Opposition: Winckelmann, Mengs, Heinse,* edited by H. Pfotenhauer et al., 9–50. Bibliothek der Kunstliteratur, no. 2. Frankfurt am Main.

Zedler, Johann Heinrich, ed. 1732–54. *Grosses vollständiges Universal-Lexicon aller Wissenschaften und Künste, welche bishero durch menschlichen Verstand und Witz erfunden und verbessert worden.* 64 vols. Halle and Leipzig.

Secondary Sources

Abel, Wilhelm. 1981. Massenarmut und Hungerkrisen in Deutschland im letzten Drittel des 18. Jahrhunderts. In *Das Pädagogische Jahrhundert,* edited by U. Herrmann, 29–52. Weinheim and Basel.

Ahrbeck, Rosemarie. 1976. Der Beitrag Basedows und des Philanthropismus zur Herausbildung des bürgerlich-humanistischen Ideals der allseitig und harmonisch entwickelten Persönlichkeit. *Jahrbuch für Erziehungs- und Schulgeschichte* 16:33–43.

Alfter, Dieter. 1986. *Die Geschichte des Augsburger Kabinettschranks.* Schwäbische Geschichtsquellen und Forschungen, no. 15. Augsburg.

Alt, Robert. 1965–66. *Bilderatlas zur Schul- und Erziehungsgeschichte.* 2 vols. Berlin.

Ariès, Philippe. 1962. *Centuries of Childhood: A Social History of Family Life.* New York.

Assmann, Aleida. 1993. *Arbeit am nationalen Gedächtnis: Eine kurze Geschichte der deutschen Bildungsidee.* Edition Pandora, no. 14. Frankfurt am Main, New York, and Paris.

Assmann, Aleida, and Dietrich Harth, eds. 1991. *Mnemosyne: Formen und Funktionen der kulturellen Erinnerung.* Frankfurt am Main.

Bachmann-Medick, Doris. 1989. *Die ästhetische Ordnung des Handelns: Moralphilosophie und Ästhetik in der Popularphilosophie des 18. Jahrhunderts.* Stuttgart.

Ballauff, Theodor, and Klaus Schaller. 1969–73. *Pädagogik: Eine Geschichte der Bildung und Erziehung.* 3 vols. Freiburg and Munich.

Barta, Ilsebill. 1987. Der disziplinierte Körper: Bürgerliche Körpersprache und ihre geschlechtsspezifische Differenzierung am Ende des 18. Jahrhunderts. In *Frauen, Bilder, Männer, Mythen,* edited by I. Barta et al., 84–106. Berlin.

Barta-Fliedl, Ilsebill, and Christoph Geissmar, eds. 1992. *Die Beredsamkeit des Leibes: Zur Körpersprache in der Kunst.* Veröffentlichungen der Albertina, no. 31. Salzburg and Vienna. Exhibition catalog.

Barthes, Roland. 1964. Images, raison, déraison. Preface to *Encyclopédie ou Dictionnaire raisonnée des sciences: L'univers de l'Encyclopédie: Les 135 plus belles planches de l'Encyclopédie de Diderot et d'Alembert,* edited by R. Barthes, 9–16, plate on p. 24. Paris.

Baudrillard, Jean. 1968. *Le Système des objets.* Paris.

Bauer, Michael. 1982. Christoph Weigel (1654–1725), Kupferstecher und Kunsthändler in Augsburg und Nürnberg. *Archiv für Geschichte des Buchwesens* 23:693–1186.

Beise, Arnd. 1992. "Meine scandaleusen Exkursionen über den Hogarth": Lichtenbergs Erklärungen zu Hogarths moralischen Kupferstichen. In *Georg Christoph Lichtenberg (1742–1799): Wagnis der Aufklärung,* compiled by U. Joost et al., 239–59. Munich and Vienna. Exhibition catalog.

Belting, Hans. 1991. *Bild und Kult: Eine Geschichte des Bildes vor dem Zeitalter der Kunst.* 2d ed. Munich.

Berg, Christa. 1983. Ansätze zu einer Sozialgeschichte des Spiels. *Zeitschrift für Pädagogik* 30:735–53.

Berg, Gunter. 1964–66. Die Selbstverlagsidee bei deutschen Autoren im 18. Jahrhundert. *Archiv für Geschichte des Buchwesens* 6:1371–96.

Berns, Jörg Jochen. 1993. Umrüstung der Mnemotechnik im Kontext von Reformation und Gutenbergs Erfindung. In *Ars memorativa: Zur kulturgeschichtlichen Bedeutung der Gedächtniskunst, 1400–1750,* edited by J. J. Berns and W. Neuber, 35–72. Tübingen.

Berns, Jörg Jochen, and Wolfgang Neuber, eds. 1993. *Ars memorativa: Zur kulturgeschichtlichen Bedeutung der Gedächtniskunst, 1400–1750.* Frühe Neuzeit, no. 15. Tübingen.

Beyer, Andreas, ed. 1992. *Die Lesbarkeit der Kunst: Zur Geistes-Gegenwart der Ikonologie.* Kleine Kulturwissenschaftliche Bibliothek, no. 37. Berlin.

Bijker, Wiebe E., et al., eds. 1987. *The Social Construction of Technological Systems:*

New Directions in the Sociology and History of Technology. Cambridge and London.

Bilstein, Johannes. 1992. Bilder für die Gestaltung des Menschen. *Neue Sammlung* 32:110–33.

Bilz, Rudolf. 1974. *Studien über Angst und Schmerz: Paläoanthropologie.* Vols. 1 and 2. Frankfurt am Main.

Blankertz, Herwig. 1981. Die utilitaristische Berufsbildungstheorie der Aufklärungspädagogik. In *Das pädagogische Jahrhundert,* edited by U. Herrmann, 247–70. Weinheim and Basel.

Bock, Friedrich. 1950. Georg Andreas Will: Ein Lebensbild aus der Spätzeit der Universität Altdorf. *Mitteilungen des Vereins für Geschichte der Stadt Nürnberg* 41:404–27.

Bogeng, Gustav Adolf Erich. 1930–41. *Geschichte der Buchdruckerkunst.* 2 vols. Hellerau and Berlin.

Böhme, Günther, and Heinz-Elmar Tenorth. 1990. *Einführung in die historische Pädagogik.* Darmstadt.

Böhme, Hartmut. 1989. Der sprechende Leib: Die Semiotiken des Körpers am Ende des 18. Jahrhunderts und ihre hermetische Tradition. In *Transfigurationen des Körpers,* edited by D. Kamper and C. Wulf, 144–81. Berlin.

Böhme, Hartmut, and Gernot Böhme. 1983. *Das Andere der Vernunft: Zur Entwicklung von Rationalitätsstrukturen am Beispiel Kants.* Frankfurt am Main.

Bolzoni, Lina. 1994. Das Sammeln und die ars memoriae. In *Macrocosmos in Microcosmo,* edited by A. Grote, 129–68. Opladen.

Bourdieu, Pierre. 1977. *Outline of a Theory of Practice.* Cambridge.

Braudel, Fernand. 1979. *Civilisation matérielle, économie, et capitalisme, XVième au XVIIIième siècle: Les structures du quotidien: Le possible et l'impossible.* Vol. 2, *Les Jeux de l'échange.* Paris.

——. 1981. *The Structures of Everyday Life.* London.

Braungart, Wolfgang. 1989. *Die Kunst der Utopie: Vom Späthumanismus zur frühen Aufklärung.* Stuttgart.

Bravo-Villasente, Carmen. 1977. *Weltgeschichte der Kinder- und Jugendliteratur: Versuch einer Gesamtdarstellung.* Hannover.

Bredekamp, Horst. 1982. Antikensehnsucht und Maschinenglauben. In *Forschungen zur Villa Albani: Antike Kunst und die Epoche der Aufklärung,* edited by H. Beck and P. C. Bol, 507–59. Frankfurter Forschungen zur Kunst, no. 10. Berlin.

——. 1993. *Antikensehnsucht und Maschinenglauben: Die Geschichte der Kunstkammer und die Zukunft der Kunstgeschichte.* Kleine Kulturwissenschaftliche Bibliothek, no. 41. Berlin.

Bremmer, Jan, and Herman Roodenburg, eds. 1992. *A Cultural History of Gesture.* Ithaca, N.Y.

Bruchmüller, Wilhelm. 1909. *Der Leipziger Student, 1409–1909.* Leipzig.

Bruford, Walter H. [1936] 1975. *Die gesellschaftlichen Grundlagen der Goethezeit.* Frankfurt am Main, Berlin, and Vienna.

Brüggemann, Theodor. 1979. Johann Siegmund Stoy. In *Lexikon der Kinder- und Jugendliteratur,* edited by K. Doderer, 3:472–75. Weinheim and Basel.

Brüggemann, Theodor, and Hans-Heino Ewers. 1982–91. *Handbuch zur Geschichte der Kinder- und Jugendliteratur.* 3 vols. Stuttgart.

Die Buchillustration im 18. Jahrhundert. 1980. Edited by Arbeitsstelle 18. Jahrhundert. Beiträge zur Geschichte der Literatur und Kunst im 18. Jahrhundert, no. 4. Heidelberg.

Busch, Werner. 1977. *Nachahmung als bürgerliches Kunstprinzip: Ikonographische Zitate bei Hogarth und in seiner Nachfolge.* Studien zur Kunstgeschichte, no. 7. Hildesheim.

———. 1984. Die Akademie zwischen autonomer Zeichnung und Handwerksdesign: Zur Auffassung der Linie und der Zeichen im 18. Jahrhundert. In *Ideal und Wirklichkeit der Bildenden Kunst im späten 18. Jahrhundert,* edited by H. Beck et al., 177–92. Berlin.

———. 1993. *Das sentimentalische Bild: Die Krise der Kunst im 18. Jahrhundert und die Geburt der Moderne.* Munich.

Cahn, Michael. 1987. Raritätenkabinette und Florilegien: Vom Wissen des Sammlers und einer nervösen Lektüre. *Tumult: Zeitschrift für Verkehrswissenschaft* 9:49–62.

Darnton, Robert. 1979. *The Business of Enlightenment: A Publishing History of the "Encyclopédie," 1775–1800.* Cambridge, Mass.

———. 1984. *The Great Cat Massacre and Other Episodes in French Cultural History.* London.

Daston, Lorraine J. 1988. The factual sensibility: Reviews on artifact and experiment. *ISIS* 79:452–70.

Davis, Natalie Zemon. 1983. *The Return of Martin Guerre.* Cambridge, Mass. Original French edition 1982.

Deutsch, Angela. 1995. Die Welt im Kleinen: Die Praxis des Universal-Sammelns im 16. Jahrhundert, eine historisch-psychologische Studie. Master's thesis, University of Vienna.

Dickerhof, Harald. 1982. Gelehrte Gesellschaften, Akademien, Ordensstudien, und Universitäten. *Zeitschrift für bayerische Landesgeschichte* 45:37–66.

Dierks, Margarete. 1965. *Vom Bilderbuch zum Arbeitsbuch.* Reutlingen.

Dierse, Ullrich. 1977. *Enzyklopädie: Zur Geschichte eines philosophischen und wissenschaftstheoretischen Begriffs.* Archiv für Begriffsgeschichte, suppl. no. 2. Bonn.

Dilg, Peter. 1994. Apotheker als Sammler. In *Macrocosmos in Microcosmo,* edited by A. Grote, 453–74. Opladen.

Doderer, Klaus, ed. 1975–82. *Lexikon der Kinder- und Jugendliteratur.* 3 vols. Weinheim and Basel.

Doderer, Klaus, and Helmut Müller, eds. 1973. *Das Bilderbuch: Geschichte und Entwicklung des Bilderbuchs in Deutschland von den Anfängen bis zur Gegenwart.* Weinheim and Basel.

Dolch, Josef. 1982. *Lehrplan des Abendlandes: Zweieinhalb Jahrtausende seiner Geschichte.* Reprint of 3d ed. of 1971. Darmstadt. Original edition 1959.

Dressen, Wolfgang. 1982. *Die pädagogische Maschine: Zur Geschichte des industrialisierten Bewußtseins in Preußen/Deutschland.* Frankfurt am Main.

Duden, Barbara. 1977. Das schöne Eigentum: Zur Herausbildung des bürgerlichen Frauenbildes an der Wende vom 18. zum 19. Jahrhundert. *Kursbuch* 47:125–40.

Dülmen, Richard van. 1973. Zum Strukturwandel der Aufklärung in Bayern. *Zeitschrift für bayerische Landesgeschichte* 36:662–79.

——. 1986. *Die Gesellschaft der Aufklärer: Zur bürgerlichen Emanzipation und aufklärerischen Kultur in Deutschland.* Frankfurt am Main.

Duncker, Ludwig. 1990. Die Kultur des Sammelns und ihre pädagogische Bedeutung. *Neue Sammlung* 30:449–65.

Dyhrenfurth-Graebsch, Irene. 1951. *Geschichte des deutschen Jugendbuches.* Hamburg.

Eichberg, Henning. 1981. Stoppuhr, Reck, und Halle: Zur Technisierung der Leibesübungen im 18. und frühen 19. Jahrhundert. In *Technologischer Wandel im 18. Jahrhundert,* edited by U. Troitzsch, 155–74. Wolfenbütteler Forschungen, no. 14. Wolfenbüttel.

Elias, Norbert. [1939] 1976. *Über den Prozeß der Zivilisation: Soziogenetische und psychogenetische Untersuchungen.* 2 vols. Frankfurt am Main.

Enklaar, Jatti. 1980. Buch und Buchillustration: Mathias Claudius' Wandsbecker Bote. In *Die Buchillustration im 18. Jahrhundert,* edited by Arbeitsstelle 18. Jahrhundert, 26–43. Beiträge zur Geschichte der Literatur und Kunst des 18. Jahrhunderts, no. 4. Heidelberg.

Ewers, Hans-Heino, ed. 1980. *Kinder- und Jugendliteratur der Aufklärung: Eine Textsammlung.* Stuttgart.

Eybl, Franz M., et al., eds. 1995. *Enzyklopädien in der Frühen Neuzeit: Beiträge zu ihrer Erforschung.* Tübingen.

Fatke, Reinhard, and Andreas Flitner. 1983. Was Kinder sammeln: Beobachtungen und Überlegungen aus pädagogischer Sicht. *Neue Sammlung* 23:600–611.

Fickert, Wilhelm. 1989. *Geldwesen, Kaufkraft, und Maßeinheiten im Bereich des Fürstentums Kulmbach-Bayreuth.* Nuremberg.

Findlen, Paula. 1989. The museum: Its classical etymology and Renaissance genealogy. *Journal of the History of Collections* 1:59–78.

——. 1994. *Possessing Nature: Museums, Collecting, and Scientific Culture in Early Modern Italy.* Berkeley.

Fischer-Homberger, Esther. 1970. *Hypochondrie: Melancholie bis Neurose: Krankheiten und Zustandsbilder.* Bern.

Florey, Ernst. 1993. Memoria: Geschichte der Konzepte über die Natur des Gedächtnisses. In *Das Gehirn: Organ der Seele? Zur Ideengeschichte der Neurobiologie,* edited by E. Florey and O. Breidbach, 151–215. Berlin.

Focke, Rudolf, ed. 1901. *Chodowiecki und Lichtenberg: Daniel Chodowiecki's Monats-*

kupfer zum "Göttinger Taschenkalender" nebst Georg Christoph Lichtenberg's Erklärungen. Leipzig.

Foucault, Michel. 1966. *Les Mots et les choses: Une archéologie des sciences humaines.* Paris.

——. 1990. *Naissance de la clinique.* 2d ed. Paris. Original edition 1963.

Fraas, Hans-Jürgen. 1988. Katechismus. In *Theologische Realenzyklopädie,* edited by G. Müller, 17:710–22. Berlin and New York.

Fricke, Wilhelm. 1886. *Grundriss der Geschichte deutscher Jugendliteratur: Ein Beitrag zur deutschen Literaturgeschichte und ein Handbuch für Eltern und Erzieher.* Minden.

Friedell, Egon. [1927–31] 1989. *Kulturgeschichte der Neuzeit. Die Krisis der europäischen Seele von der schwarzen Pest bis zum ersten Weltkrieg.* Munich.

Gallwitz, Klaus, and Margret Stuffmann, eds. 1978. *Bürgerliches Leben im 18. Jahrhundert: Daniel Chodowiecki, 1726–1801, Zeichnungen und Druckgraphik.* Compiled by P. Märker. Frankfurt am Main. Exhibition catalog.

Garber, Daniel. 1992. *Descartes' Metaphysical Physics.* Chicago.

Gerlach, Peter. 1984. Über das mittlere Maß, oder Der bürgerliche Kanon. In *Ideal und Wirklichkeit der bildenden Kunst im späten 18. Jahrhundert,* edited by H. Beck and P. C. Bol, 45–74. Frankfurter Forschungen zur Kunst, no. 11. Berlin.

Giedion, Sigfried. 1948. *Mechanization Takes Command: A Contribution to Anonymous History.* New York.

Giesecke, Michael. 1991. *Der Buchdruck in der frühen Neuzeit: Eine historische Fallstudie über die Durchsetzung neuer Informations- und Kommunikationstechnologien.* Frankfurt am Main.

Ginzburg, Carlo. 1983. *Spurensicherungen: Über verborgene Geschichte, Kunst, und soziales Gedächtnis.* Berlin.

Göbels, Hubert. 1965. J. S. Stoy: Silhouetten der berühmten Pädagogen. *Lehren und lernen* 1:42–45.

——. 1980. *Hundert alte Kinderbücher aus Barock und Aufklärung: Eine illustrierte Bibliographie.* Die bibliophilen Taschenbücher, no. 196. Dortmund.

Göhring, Ludwig. 1904. *Die Anfänge der deutschen Jugendliteratur im 18. Jahrhundert.* Nuremberg.

Goldfriedrich, Johann. 1909. *Geschichte des deutschen Buchhandels vom Beginn der klassischen Literaturperiode bis zum Beginn der Fremdherrschaft, 1740–1804.* Geschichte des deutschen Buchhandels, no. 3. Leipzig.

Gombrich, Ernst Hans. 1970. *Aby Warburg: An Intellectual Biography.* London.

Gooding, David, Trevor Pinch, and Simon Schaffer, eds. 1989. *The Uses of Experiment: Studies in the Natural Sciences.* Cambridge.

Göpfert, Herbert G., et al., eds. 1977. *Buch- und Verlagswesen im 18. und 19. Jahrhundert.* Studien zur Geschichte der Kulturbeziehungen in Mittel- und Osteuropa, no. 4. Berlin.

Grenz, Dagmar, ed. 1986. *Aufklärung und Kinderbuch: Studien zur Kinder- und Jugendliteratur des 18. Jahrhunderts.* Pinneberg.

Grote, Andreas, ed. 1994. *Macrocosmos in Microcosmo: Die Welt in der Stube: Zur Geschichte des Sammelns 1450 bis 1800.* Berliner Schriften zur Museumskunde, no. 10. Opladen.

Habermas, Jürgen. 1962. *Strukturwandel und Öffentlichkeit: Untersuchungen zu einer Kategorie der bürgerlichen Gesellschaft.* Politica, no. 4. Neuwied and Berlin.

Hagner, Michael. 1990. Die Entfaltung der cartesischen "Mechanik des Sehens" und ihre Grenzen. *Sudhoffs Archiv* 74:1–22.

Haltern, Utz. 1985. *Bürgerliche Gesellschaft: Sozialtheoretische und sozialhistorische Aspekte.* Erträge der Forschung, no. 227. Darmstadt.

Hammermayer, Ludwig. 1976. Akademiebewegung und Wissenschaftsorganisation: Formen, Tendenzen, und Wandel in Europa während der zweiten Hälfte des 18. Jahrhunderts. In *Wissenschaftspolitik in Mittel- und Osteuropa: Wissenschaftliche Gesellschaften, Akademien, und Hochschulen im 18. und beginnenden 19. Jahrhundert,* edited by E. Amburger et al., 1–84. Studien zur Geschichte der Kulturbeziehungen in Mittel- und Osteuropa, no. 3. Berlin.

Haniel, Joachim. 1962. Kirchenhoheit und Kirchenregiment des Nürnberger Rates in den letzten Jahren der Reichsfreiheit und deren Übernahme durch Bayern. *Mitteilungen des Vereins für Geschichte der Stadt Nürnberg* 51:316–432.

Harms, Wolfgang, ed. 1990. *Text und Bild, Bild und Text.* Germanistisches Symposion der Deutschen Forschungsgemeinschaft, no. 11. Stuttgart.

Hartmann, Andreas. 1989. Zur Geschichte der Gedächtnissysteme. *Beiträge zur Volkskunde in Niedersachsen* 5:47–62.

Hartmann, Fritz, and Rudolf Vierhaus. 1977. *Der Akademiegedanke im 17. und 18. Jahrhundert.* Wolfenbütteler Forschungen, no. 3. Bremen and Wolfenbüttel.

Haskell, Francis. 1993. *History and Its Images: Art and the Interpretation of the Past.* New Haven and London.

Hauger, Harriet. 1996. Samuel Quiccheberg, oder Der Anfang der Museumslehre in Deutschland. Ph.D. dissertation, Humboldt University, Berlin.

Hausenstein, Wilhelm. 1958. *Rokoko: Französische und deutsche Illustratoren des 18. Jahrhunderts.* Munich.

Hauser, Andrea. 1994. *Dinge des Alltags: Studien zur historischen Sachkultur eines schwäbischen Dorfes.* Untersuchungen des Ludwig-Uhland-Instituts der Universität Tübingen, no. 82. Tübingen.

Hauswedell, Ernst Ludwig, and Christian Voigt. 1977. *Buchkunst und Literatur in Deutschland, 1750–1850.* 2 vols. Hamburg.

Henkel, Arthur, and Albrecht Schöne, eds. 1967. *Emblemata: Handbuch zur Sinnbildkunst des 16. und 17. Jahrhunderts.* Stuttgart.

Henningsen, Jürgen. 1966. "Enzyklopädie": Zur Sprach- und Bedeutungsgeschichte eines pädagogischen Begriffs. *Archiv für Begriffsgeschichte* 10:271–362.

Herklotz, Ingo. 1994. Neue Literatur zur Sammlungsgeschichte. *Kunstchronik* 47:117–35.

Herrlinger, Robert. 1967. *Geschichte der medizinischen Abbildung: Von der Antike bis um 1600.* Vol. 1. Munich.

Herrmann, Ulrich, ed. 1981. *"Das pädagogische Jahrhundert": Volksaufklärung und Erziehung zur Armut im 18. Jahrhundert in Deutschland.* Geschichte des Erziehungs- und Bildungswesens in Deutschland, no. 1. Weinheim and Basel.

——, ed. 1982. *Die Bildung des Bürgers: Die Formierung der bürgerlichen Gesellschaft und die Gebildeten im 18. Jahrhundert.* Geschichte des Erziehungs- und Bildungswesens in Deutschland, no. 2. Weinheim and Basel.

——. 1993. *Aufklärung und Erziehung: Studien zur Funktion der Erziehung im Konstitutionsprozeß der bürgerlichen Gesellschaft im 18. und frühen 19. Jahrhundert in Deutschland.* Weinheim.

Heyde, Johannes Erich. 1935. *Technik des wissenschaftlichen Arbeitens: Eine Anleitung, besonders für Studierende, mit ausführlichem Schriftenverzeichnis.* 5th ed. Berlin.

Hiller, Helmut. 1966. *Zur Sozialgeschichte von Buch und Buchhandel.* Bonner Beiträge zur Bibliotheks- und Bücherkunde, no. 13. Bonn.

Hobrecker, Karl. 1924. *Alte vergessene Kinderbücher.* Berlin.

Hoffmann, Edith. 1934. *Die Darstellung des Bürgers in der deutschen Malerei des 18. Jahrhunderts.* Berlin.

Honegger, Claudia, ed. 1977. *M. Bloch, F. Braudel, L. Febvre u.a. Schrift und Materie der Geschichte: Vorschläge zur systematischen Aneignung historischer Prozesse.* Frankfurt am Main.

Hooper-Greenhill, Eilean. 1992. *Museums and the Shaping of Knowledge.* London and New York.

Hruby, Ingrid. 1986. Pansophie und Polymathie: Überlegungen zum Sachbuch für Kinder im 18. Jahrhundert. In *Aufklärung und Kinderbuch,* edited by D. Grenz, 153–82. Pinneberg.

Hurrelmann, Bettina. 1974. *Jugendliteratur und Bürgerlichkeit: Soziale Erziehung in der Jugendliteratur der Aufklärung am Beispiel von Christian Felix Weißes Kinderfreund, 1776–1782.* Informationen zur Sprach- und Literaturdidaktik, no. 5. Paderborn.

——. 1982. Erziehung zur Bürgerlichkeit in der Jugendliteratur der Aufklärung: Am Beispiel von Christian Felix Weißes "Kinderfreund," 1776–1782, gezeigt. In *Die Bildung des Bürgers,* edited by U. Herrmann, 194–223. Weinheim and Basel.

Immerwahr, Raymond. 1978. Diderot, Herder, and the dichotomy of touch and sight. *Seminar* 14:84–96.

Impey, Oliver, and Arthur MacGregor, eds. 1985. *The Origins of Museums: The Cabinet of Curiosities in Sixteenth- and Seventeenth-Century Europe.* Oxford.

Jahn, Ilse. 1994. Sammlungen: Aneignung und Verfügbarkeit. In *Macrocosmos in Microcosmo,* edited by A. Grote, 475–500. Opladen.

Jahn, Ilse and Konrad Senglaub. 1978. *Carl von Linné.* Biographien hervorragender Naturwissenschaftler, Techniker, und Mediziner, no. 35. Leipzig.

Jay, Martin. 1988. Scopic regimes of modernity. In *Vision and Visuality,* edited by Hal Foster, 3–23. Dia Art Foundation Discussions in Contemporary Culture, no. 2. Seattle.

Jentzsch, Rudolf. 1912. *Deutsch-lateinischer Büchermarkt nach den Leipziger Ostermeßkatalogen von 1740, 1770, und 1800 in seiner Gliederung und Wandlung.* Beiträge zur Kultur- und Universalgeschichte, no. 22. Leipzig.

Johns, Adrian. 1998. *The Nature of Book: Print and Knowledge in the Making.* Chicago and London.

Junker, Almut, and Eva Stille, eds. 1984. *Spielen und lernen: Spielzeug und Kinderleben in Frankfurt, 1750–1930.* Kleine Schriften des Historischen Museums Frankfurt, no. 22. Frankfurt am Main. Exhibition catalog.

Keck, Rudolf W. 1986. Die Entdeckung des Bildes durch die Pädagogik. In *Bildungsgeschichte als Sozialgeschichte: Festschrift zum 60. Geburtstag von Franz Poeggeler,* edited by H. Kanz, 81–124. Erziehungsphilosophie, no. 8. Frankfurt am Main.

———. 1991. Die Entdeckung des Bildes in der erziehungstheoretischen Forschung. In *Bild und Bildung: Ikonologische Interpretationen vormoderner Dokumente von Erziehung und Bildung,* edited by C. Rittelmeyer and E. Wiersing, 23–52. Wolfenbütteler Forschungen, no. 49. Wiesbaden.

Kellenbenz, Hermann. 1977. *Deutsche Wirtschaftsgeschichte.* Vol. 1, *Von den Anfängen bis zum Ende des 18. Jahrhunderts.* Munich.

Kemp, Wolfgang. 1975. Die Beredsamkeit des Leibes: Körpersprache als künstlerisches und gesellschaftliches Problem der bürgerlichen Emanzipation. *Städel-Jahrbuch,* n.s., 5:111–34.

Kersting, Christa. 1992. *Die Genese der Pädagogik im 18. Jahrhundert: Campes "Allgemeine Revision" im Kontext der neuzeitlichen Wissenschaft.* Weinheim.

Kirchhoff, R. 1974. Gebärde, Gebärdensprache. In *Historisches Wörterbuch der Philosophie,* edited by J. Ritter, 3:30–31. Basel.

Kleinau, Elke, and Claudia Opitz, eds. 1996. *Geschichte der Mädchen- und Frauenbildung.* Vol. 1, *Vom Mittelalter bis zur Aufklärung.* Frankfurt am Main and New York.

Klemm, Friedrich. 1973. *Geschichte der naturwissenschaftlichen und technischen Museen.* Deutsches Museum, Abhandlungen und Berichte, no. 41.2. Munich.

Köberle, Sophie. [1924] 1972. *Jugendliteratur zur Zeit der Aufklärung.* Internationale Untersuchungen zur Kinder- und Jugendliteratur, no. 4. Weinheim and Basel.

Köhler, Woldemar. 1896. *Zur Entwicklungsgeschichte des Buchgewerbes von der Erfindung der Buchdruckerkunst bis zur Gegenwart.* Gera-Untermhaus.

Kossmann, Bernhard. 1967–69. Deutsche Universallexika des 18. Jahrhunderts: Ihr Wesen und Informationswert, dargestellt am Beispiel der Werke von Jablonski und Zedler. *Archiv für Geschichte des Buchwesens* 9:1553–96.

Köster, Hermann Leopold. [1906] 1972. *Geschichte der deutschen Jugendliteratur.* Munich-Pullach and Berlin.

Kraus, Andreas. 1977. Die Bedeutung der deutschen Akademien des 18. Jahrhunderts für die historische und naturwissenschaftliche Forschung. In *Der Akademiegedanke im 17. und 18. Jahrhundert,* edited by F. Hartmann and R. Vierhaus, 139–70. Bremen and Wolfenbüttel.

Krause, Horst. 1988. Die allgemeinbildende Dimension im *Elementarwerk* Johann Bernhard Basedows (1724–1790). In *Zwischen Renaissance und Aufklärung: Beiträge der interdisziplinären Arbeitsgruppe Frühe Neuzeit der Universität Osnabrück /Vechta,* edited by K. Garber and W. Kürschner, 273–98. Amsterdam.

Krebs, Margarete. 1929. *Elementarwerke aus der Zeit des Philanthropismus.* Coburg.

Kris, Ernst, and Otto Kurz. [1934] 1980. *Die Legende vom Künstler: Ein geschichtlicher Versuch.* Frankfurt am Main.

Krolzik, Udo. 1980. Das physikotheologische Naturverständnis und sein Einfluß auf das naturwissenschaftliche Denken im 18. Jahrhundert. *Medizinhistorisches Journal* 15:90–102.

Kubler, George. 1963. *The Shape of Time: Remarks on the History of Things.* New Haven.

Kunze, Horst. 1965. *Schatzbehalter: Vom Besten aus der älteren deutschen Kinderliteratur.* Hanau am Main.

Kunzle, David. 1973. *The Early Comic Strip: Narrative Strips and Picture Stories in the European Broadsheet from c. 1450 to 1825.* Berkeley.

Kutra, G. 1906. Abraham in der bildenden Kunst. *Ost und West* 6:701–24.

Lanckorónska, Maria, and Richard Oehler. 1932–34. *Die Buchillustration des 18. Jahrhunderts in Deutschland, Österreich, und der Schweiz.* 3 vols. Frankfurt am Main.

Lanz, Jakob. 1971. Affekt. In *Historisches Wörterbuch der Philosophie,* edited by J. Ritter, 1:93–100. Basel.

Latour, Bruno. 1987. *Science in Action.* Cambridge, Mass.

———. 1990. Drawing things together. In *Representation in Scientific Practice,* edited by M. Lynch and S. Woolgar, 19–68. Cambridge and London.

Ledderhose, Maria. 1982. Daniel Chodowiecki und die Pädagogik im 18. Jahrhundert. Ph.D. dissertation, University of Cologne.

Leder, Klaus. 1965. *Universität Altdorf: Zur Theologie der Aufklärung in Franken: Die theologische Fakultät in Altdorf, 1750–1809.* Schriftenreihe der Altnürnberger Landschaft, no. 14. Nuremberg.

Lehmann, Hannelore. 1971. Paul Jacob Marperger (1656–1730), ein vergessener Ökonom der deutschen Frühaufklärung. *Jahrbuch für Wirtschaftsgeschichte* 4:125–57.

Lepenies, Wolf. 1976. *Das Ende der Naturgeschichte: Wandel kultureller Selbstverständlichkeiten in den Wissenschaften des 18. und 19. Jahrhunderts.* Munich and Vienna.

Leroi-Gourhan, André. 1964–65. *Le Geste et la parole.* Paris.

Liebenwein, Wolfgang. 1977. *Studiolo: Die Entstehung eines Raumtyps und seine Entwicklung bis um 1600.* Frankfurter Forschungen zur Kunst, no. 6. Berlin.

Louis, Eleonora. 1992. Der beredte Leib: Bilder aus der Sammlung Lavater. In *Die Beredsamkeit des Leibes,* edited by I. Barta-Fliedl and C. Geissmar, 113–55. Salzburg and Vienna. Exhibition catalog.

Maassen, Nikolaus, ed. 1960. *Geschichte der Mittel- und Realschulpädagogik.* Vol. 1, *Von den Anfängen bis Ende des 19. Jahrhunderts,* by W. Schöler. Berlin.

MacGregor, Arthur. 1994. Die besonderen Eigenschaften der "Kunstkammer." In *Macrocosmos in Microcosmo,* edited by A. Grote, 61–106. Opladen.

Martens, Wolfgang. 1968. *Die Botschaft der Tugend: Die Aufklärung im Spiegel der deutschen moralischen Wochenschriften.* Stuttgart.

Matuszak, Juliane. 1967. *Das Speculum exemplorum als Quelle volkstümlicher Glaubensvorstellungen des Spätmittelalters.* Quellen und Studien zur Volkskunde, no. 8. Siegburg.

Medicus, Emil Friedrich Heinrich. 1863. *Geschichte der evangelischen Kirche im Königreiche Bayern diesseits des Rheins, nach gedruckten und teilweise auch ungedruckten Quellen zunächst für praktische Geistliche und sonstige gebildete Leser bearbeitet.* Erlangen.

Meier, Hans Jakob. 1994. *Die Buchillustration des 18. Jahrhunderts in Deutschland und die Auflösung des überlieferten Historienbildes.* Kunstwissenschaftliche Studien, no. 60. Munich.

Meinel, Christoph. 1995. Enzyklopädie der Welt und Verzettelung des Wissens: Aporien der Empirie bei Joachim Jungius. In *Enzyklopädien der Frühen Neuzeit,* edited by F. M. Eybl et al., 162–87. Tübingen.

Menze, Clemens. 1966. Die Hinwendung der deutschen Pädagogik zu den Erfahrungswissenschaften vom Menschen: Eine geschichtliche Betrachtung. *Zur Bedeutung der Empirie für die Pädagogik als Wissenschaft: Neue Folge der Ergänzungshefte zur Vierteljahresschrift für wissenschaftliche Pädagogik* 5:26–52.

Merkle, Siegbert Ernst. 1983. *Die historische Dimension des Prinzips der Anschauung: Historische Fundierung und Klärung terminologischer Tendenzen des didaktischen Prinzips der Anschauung von Aristoteles bis Pestalozzi.* Europäische Hochschulschriften, ser. 11, no. 151. Frankfurt am Main and Bern.

Milde, W. 1989. Dedikationsbild. In *Lexikon des gesamten Buchwesens,* edited by S. Corsten et al., 2d ed., 2:237–38. Stuttgart.

Miller, Daniel, ed. 1998. *Material Cultures: Why Some Things Matter.* Chicago.

Mohr, Rudolf. 1982. Erbauungsliteratur II. In *Theologische Realenzyklopädie,* edited by G. Krause and G. Müller, 10:43–50. Berlin and New York.

Mollenhauer, Klaus. 1983a. Streifzug durch fremdes Terrain: Interpretation eines Bildes aus dem Quattrocento in bildungstheoretischer Absicht. *Zeitschrift für Pädagogik* 2:173–94.

———. 1983b. *Vergessene Zusammenhänge: Über Kultur und Erziehung.* Munich.

Morgan, Michael J. 1977. *Molyneux's Question: Vision, Touch, and the Philosophy of Perception.* Cambridge.

Muensterberger, Werner. 1994. *Collecting: An Unruly Passion: Psychological Perspectives.* Princeton.

Müller, Thomas. 1995. Die Kunst- und Naturalienkammer der Franckeschen Stiftungen. *Museumsnachrichten Sachsen-Anhalt* 2:10–17.

Müller-Dietz, Heinz 1989. Anatomische Präparate in der Petersburger "Kunstkammer." *Zentralblatt für allgemeine Pathologie und pathologische Anatomie* 135:757–67.

Müller-Rolli, Sebastian. 1989. Bilderwelt, Spiegelwelt: Über Bilder und deren Bildungsgehalt. In *Bildung, Glaube, Aufklärung: Zur Wiedergewinnung des Bildungsbegriffs in Pädagogik und Theologie,* edited by R. Preul, 37–60. Gütersloh.

Münch, Paul, ed. 1984. *Ordnung, Fleiß, und Sparsamkeit: Texte und Dokumente zur Entstehung der "bürgerlichen Tugenden."* Munich.

———. 1988. Grundwerte der frühneuzeitlichen Ständegesellschaft? Aufriß einer vernachlässigten Thematik. In *Ständische Gesellschaft und soziale Mobilität,* edited by W. Schulze, 53–72. Munich.

Müsch, Irmgard. 2000. *Geheiligte Naturwissenschaft: Die Kupfer-Bibel des Johann Jakob Scheuchzer.* Göttingen.

Niedermeier, Michael. 1996. Campe als Direktor des Dessauer Philanthropins. In *Visionäre Lebensklugheit: Joachim Heinrich Campe in seiner Zeit (1746–1818),* edited by H. Schmitt, 45–65. Wiesbaden. Exhibition catalog.

Nipperdey, Thomas. 1993. *Deutsche Geschichte, 1800–1866, Bürgerwelt und starker Staat.* Munich.

Oertel, Hermann. 1977. Das Bild in Bibeldrucken vom 15. bis zum 18. Jahrhundert (Die Wolfenbütteler Bibelsammlung). *Jahrbuch der Gesellschaft für niedersächsische Kirchengeschichte* 75:9–37.

Ophir, Adi, and Steven Shapin. 1991. The place of knowledge: A methodological survey. *Science in Context* 4, no. 1:3–21.

Outram, Dorinda. 1996. New spaces in natural history. In *Cultures of Natural History,* edited by N. Jardine et al., 249–65. Cambridge.

Peil, Dietmar. 1977–78. Zur Illustrationsgeschichte von Johann Arndts "Vom wahren Christentum." *Archiv für Geschichte des Buchwesens* 18:963–1066.

Philipp, Wolfgang. 1957. *Das Werden der Aufklärung in theologiegeschichtlicher Sicht.* Forschungen zur systematischen Theologie und Religionsphilosophie, no. 3. Göttingen.

Pilz, Kurt, compiler. 1967. *Die Ausgaben des Orbis sensualium pictus: Eine Bibliographie.* Nuremberg.

Pinloche, Auguste. 1896. *Geschichte des Philanthropinismus.* Leipzig.

Pomian, Krzysztof. 1987. *Collectionneurs, connaisseurs, et curieux: Paris—Venise XVIe siècle.* Paris.

———. 1988. *Der Ursprung des Museums: Vom Sammeln.* Kleine Kulturwissenschaftliche Bibliothek, no. 9. Berlin.

——. 1994. Sammlungen: Eine historische Typologie. In *Macrocosmos in Microcosmo,* edited by A. Grote, 107–26. Opladen.

Porstmann, Walter. 1928. *Karteikunde: Das Handbuch der Karteitechnik.* Stuttgart.

Porter, Roy. 1990. *The Enlightenment.* Studies in European History. Basingstoke and London.

Pressler, Christine. 1980. *Schöne alte Kinderbücher.* Munich.

Promies, Wolfgang. 1980. Kinderliteratur im späten 18. Jahrhundert. In *Hansers Sozialgeschichte der deutschen Literatur,* edited by R. Grimminger, 3:765–831. Munich.

Raabe, Paul. 1977. Buchproduktion und Lesepublikum in Deutschland, 1770–1780. *Philobiblion* 21:2–16.

Raff, Thomas. 1994. *Die Sprache der Materialien: Anleitung zu einer Ikonologie der Werkstoffe.* Kunstwissenschaftliche Studien, no. 61. Munich.

Rammensee, Dorothea. 1961. *Bibliographie der Nürnberger Kinder- und Jugendbücher, 1522–1914.* Bamberg.

Raulff, Ulrich, ed. 1987. *Mentalitäten-Geschichte: Zur historischen Rekonstruktion geistiger Prozesse.* Berlin.

Rebhuhn, Adolf. 1925. *Pädagogisches Druckgut vergangener Jahrhunderte.* Berlin.

Reich, Detlef. 1981. Johann Siegmund Stoys "Bilder-Akademie für die Jugend." Essay at the Faculty of German Language and Literature and Its Didactics, University of Cologne.

Reicke, Emil. 1896. *Geschichte der Reichsstadt Nürnberg von dem ersten urkundlichen Nachweis ihres Bestehens bis zu ihrem Übergang an das Königreich Bayern.* Nuremberg.

Retter, Hein. 1979. *Spielzeug: Handbuch zur Geschichte und Pädagogik der Spielmittel.* Weinheim and Basel.

Reuter, Christiane. 1994. Johann Siegmund Stoys "Bilder-Akademie für die Jugend": Ein Bilder-Lehrbuch der Aufklärung. Master's thesis, University of Erlangen-Nuremberg.

Reynst, Elisabeth. 1962. *Friedrich Campe und sein Bilderbogen-Verlag zu Nürnberg: Mit einer Schilderung des Nürnberger Kunstbetriebes im 18. und in der ersten Hälfte des 19. Jahrhunderts.* Veröffentlichungen der Stadtbibliothek, no. 5. Nuremberg.

Rheinberger, Hans-Jörg. 1992. *Experiment, Differenz, Schrift: Zur Geschichte epistemischer Dinge.* Marburg.

Rheinberger, Hans-Jörg, et al., eds. 1997. *Räume des Wissens: Repräsentation, Codierung, Spur.* Berlin.

Richter, Dieter. 1987. *Das fremde Kind: Zur Entstehung der Kindheitsbilder des bürgerlichen Zeitalters.* Frankfurt am Main.

Riedel, Wolfgang. 1994. Anthropologie und Literatur in der deutschen Spätaufklärung: Skizze einer Forschungslandschaft. *Internationales Archiv für Sozialgeschichte der deutschen Literatur,* special issue no. 6:93–157.

Ries, Hans. 1982. Grundriß zu einer bibliographischen Behandlung von Illustration und optischer Erscheinungsform im historischen Kinder- und Jugendbuch. *Die Schiefertafel* 3:98–122.

Ringshausen, Gerhard. 1976. *Von der Buchillustration zum Unterrichtsmedium: Der Weg des Bildes in die Schule dargestellt am Beispiel des Religionsunterrichtes.* Studien und Dokumentationen zur deutschen Bildungsgeschichte, no. 2. Weinheim and Basel.

Roeck, Bernd. 1991. *Lebenswelt und Kultur des Bürgertums in der frühen Neuzeit.* Enzyklopädie deutscher Geschichte, no. 9. Munich.

Rosenstrauch, Hazel. 1986. Buchhandelsmanufaktur und Aufklärung: Die Reformen des Buchhändlers und Verlegers Ph. E. Reich (1717–1787): Sozialgeschichtliche Studien zur Entwicklung des literarischen Marktes. *Archiv für Geschichte des Buchwesens* 26:1–129.

Rothe, Hans, and Andrzej Ryszkiewicz, eds. 1986. *Chodowiecki und die Kunst der Aufklärung in Polen und Preußen.* Schriften des Komitees der Bundesrepublik Deutschland zur Förderung der Slawischen Studien, no. 9. Cologne and Vienna.

Rümann, Arthur. 1931. *Das deutsche illustrierte Buch des 18. Jahrhunderts.* Strasbourg.

———. 1937. *Alte deutsche Kinderbücher.* Vienna.

Rutschky, Katharina. 1988. *Schwarze Pädagogik.* Frankfurt am Main.

Sabean, David Warren. 1984. *Power in the Blood: Popular Culture and Village Discourse in Early Modern Germany.* Cambridge and New York.

Schaffer, Simon. 1983. Natural philosophy and public spectacle in the eighteenth century. *History of Science* 21:1–43.

Schaller, Klaus. 1962. *Die Pädagogik des Johann Amos Comenius und die Anfänge des pädagogischen Realismus im 17. Jahrhundert.* Pädagogische Forschungen, no. 21. Heidelberg.

Scheicher, Elisabeth. 1979. *Die Kunst- und Wunderkammern der Habsburger.* Vienna.

Schiller, Gertrud. 1966–91. *Ikonographie der christlichen Kunst.* 5 vols. Gütersloh.

Schlosser, Julius von. 1908. *Die Kunst- und Wunderkammern der Spätrenaissance.* Monographien des Kunstgewerbes, no. 11. Leipzig.

Schloz, Thomas. 1986. Alltagskultur: Sammeln. In *Enzyklopädie Erziehungswissenschaften,* edited by D. Lenzen, 3:352–56. Stuttgart.

Schmitt, Hanno, et al., eds. 1997. *Bilder als Quellen der Erziehungsgeschichte.* Bad Heilbrunn.

Schmitz, W. 1989. Dedikation. In *Lexikon des gesamten Buchwesens,* edited by S. Corsten et al., 2d ed., 2:236–37, Stuttgart.

Schöler, Walter. 1970. *Geschichte des naturwissenschaftlichen Unterrichts im 17. bis 19. Jahrhundert: Erziehungstheoretische Grundlegung und schulgeschichtliche Entwicklung.* Berlin.

Schrötter, Georg. 1908. *Die Nürnberger Malerakademie und Zeichenschule.* Würzburg.

Schug, Albert, ed. 1988. *Die Bilderwelt im Kinderbuch: Kinder- und Jugendbücher aus fünf Jahrhunderten.* Cologne. Exhibition catalog.

Schultheiss, Wolfgang Konrad. 1853–57. *Geschichte der Schulen in Nürnberg.* Issues 1–5. Nuremberg.

Shapin, Steven. 1988. The house of experiment in seventeenth-century England. *ISIS* 79:373–404.

Shapin, Steven, and Simon Schaffer. 1985. *Leviathan and the Air Pump: Hobbes, Boyle, and the Experimental Life.* Princeton.

Sibum, Heinz-Otto. 1990. *Physik aus ihrer Geschichte verstehen: Entstehung und Entwicklung naturwissenschaftlicher Denk- und Arbeitsstile in der Elektrizitätsforschung des 18. Jahrhunderts.* Wiesbaden.

———. 1995. Reworking the mechanical value of heat: Instruments of precision and gestures of accuracy in early Victorian England. *Studies in the History and Philosophy of Science* 26, no. 1:73–106.

Siemer, Stefan. 2000. *Geselligkeit und Methode: Naturgeschichtliches Sammeln im 18. Jahrhundert.* Dissertation, University of Zurich.

Simon, Matthias. 1965. *Nürnbergisches Pfarrerbuch: Die evangelisch-lutherische Geistlichkeit der Reichsstadt Nürnberg und ihrer Gebiete, 1524–1806.* Nuremberg.

Spary, Emma. 1999. Essay review: This book is a box. *Studies in the History and Philosophy of Science* 30, no. 2:355–62.

Spicer, Joaneath. 1992. The Renaissance elbow. In *A Cultural History of Gesture,* edited by J. Bremmer and H. Roodenburg, 84–128. Ithaca, N.Y.

Stach, Reinhard. 1970. Die Lernorganisation des "Denklehrzimmers": Ein methodischer Entwurf Christian Heinrich Wolkes. *Lehren und lernen* 7:558–71.

———. 1972. Das Naturalienkabinett und seine Bedeutung für die Entfaltung der anschauenden Erkenntnis. In *Essener Pädagogische Beiträge: Die Schule im Wandel der Gesellschaft,* edited by E. Luckner, 21–39. Ratingen.

———. 1974. Das Basedowsche Elementarwerk. *Pädagogica Historica* 14:458–96.

Stadtarchiv Nürnberg, ed. 1966. *Die Zeit der Aufklärung in Nürnberg, 1780–1810.* Quellen zur Geschichte und Kultur der Stadt Nürnberg, no. 6. Nuremberg. Exhibition catalog.

Stafford, Barbara Maria. 1991. *Body Criticism: Imaging the Unseen in Enlightenment Art and Medicine.* Cambridge, Mass., and London.

Starobinski, Jean. 1989. *Le Remède dans le mal: Critique et légitimation de l'artifice à l'âge des Lumiéres.* Paris.

Steiger, Rudolf. 1927–30. *Johann Jakob Scheuchzer (1672–1733) I. Werdezeit (bis 1699).* Schweizer Studien zur Geschichtswissenschaft, no. 15. Zürich-Selnau.

Storz, Jürgen. 1962. Das Naturalien- und Kunstkabinett der Franckeschen Stiftungen zu Halle an der Saale. *Wissenschaftliche Zeitschrift der Martin-Luther-Universität Halle-Wittenberg* 11:193–200.

Strobach, Erich, ed. 1978. *Alte deutsche Kinderbücher, Sammlung Strobach.* Paderborn. Exhibition catalog.

———. 1979. Johannes Buno (1617–1697), ein Zeitgenosse des Comenius: Gedanken eines Kinderarztes und Kinderbuchsammlers zum Problem des Bildes. *Die Schiefertafel* 2:72–87.

Talkenberger, Heike. 1994. Von der Illustration zur Interpretation: Das Bild als historische Quelle. *Zeitschrift für historische Forschung* 21:289–313.

te Heesen, Anke. 1996. Kinder, Kammern, Körbe: Vom Sammeln und Ordnen in einer Bildenzyklopädie der Aufklärung. *Das achtzehnte Jahrhundert* 20:150–65.

———. 1997. Verbundene Bilder: Das Tableau in den Erziehungsvorstellungen des 18. Jahrhunderts. In *Bilder als Quellen der Erziehungsgeschichte,* edited by H. Schmitt et al., 77–90. Bad Heilbrunn.

Teuscher, Adolf. 1911–12. *Joh. Heinr. Gottlieb Heusinger als Pädagoge: Eine genetische Darstellung seiner Erziehungsideen.* Pädagogisches Magazin, no. 459. Langensalza.

Thanner, Brigitte. 1987. Schweizerische Buchillustration im Zeitalter der Aufklärung am Beispiel von Johann Rudolf Schellenberg. 3 vols. Ph.D. dissertation, Ludwig-Maximilians-Universität, Munich.

Thanner, Brigitte, et al. 1987. *Johann Rudolf Schellenberg: Der Künstler und die naturwissenschaftliche Illustration im 18. Jahrhundert.* Neujahrsblätter der Stadtbibliothek Winterthur, no. 318. Winterthur.

Trefzer, Rudolf. 1989. *Die Konstruktion des bürgerlichen Menschen: Aufklärungspädagogik und Erziehung im ausgehenden 18. Jahrhundert am Beispiel der Stadt Basel.* Zürich.

Ungern-Sternberg, Wolfgang von. 1980. Schriftsteller und literarischer Markt. In *Hansers Sozialgeschichte der deutschen Literatur,* edited by R. Grimminger, 3:133–85. Munich.

Uphaus-Wehmeier, Annette. 1984. *Zum Nutzen und Vergnügen: Jugendschriften des 18. Jahrhunderts.* Dortmunder Beiträge zur Zeitungsforschung, no. 38. Munich.

Vierhaus, Rudolf, ed. 1981. *Bürger und Bürgerlichkeit im Zeitalter der Aufklärung.* Wolfenbütteler Studien zur Aufklärung, no. 7. Heidelberg.

Visser, Robert. 1990. Die Rezeption der Anthropologie Petrus Campers (1770–1850). In *Die Natur des Menschen: Probleme der physischen Anthropologie und Rassenkunde, 1750–1850,* edited by G. Mann and F. Dumont, 325–35. Soemmerring Forschungen, no. 6. Stuttgart and New York.

Vogel, Heiner. 1981. *Bilderbogen, Papiersoldat, Würfelspiel, und Lebensrad.* Würzburg.

Vogel, Martin. 1978. Deutsche Urheber- und Verlagsrechtsgeschichte zwischen 1450 und 1850: Sozial- und methodengeschichtliche Entwicklungsstufen der Rechte von Schriftsteller und Verleger. *Archiv für Geschichte des Buchwesens* 19:1–190.

Warburg, Aby M. 1992. Einleitung zum Mnemosyne-Atlas. In *Die Beredsamkeit des Leibes,* edited by I. Barta-Fliedl and C. Geissmar, 171–73. Salzburg and Vienna. Exhibition catalog.

Warncke, Carsten-Peter. 1987. *Sprechende Bilder, sichtbare Worte: Das Bildverständnis in der frühen Neuzeit.* Wolfenbütteler Forschungen, no. 33. Wiesbaden.

Warnke, Martin. 1987. Der Kopf in der Hand. In *Zauber der Medusa: Europäische Manierismen,* edited by W. Hofmann, 55–61. Vienna. Exhibition catalog.

———. 1992. Erhobenen Hauptes. In *Die Beredsamkeit des Leibes,* edited by I. Barta-Fliedl and C. Geissmar, 190–94. Salzburg and Vienna. Exhibition catalog.

Weckwerth, Alfred. 1979. Armenbibeln. In *Theologische Realenzyklopädie,* edited by G. Krause and G. Müller, 4:8–10. Berlin and New York.

Wegehaupt, Heinz. 1979. *Alte deutsche Kinderbücher: Bibliographie, 1507–1850.* Berlin.

———, ed. 1992. *Robinson und Struwwelpeter: Bücher für Kinder aus fünf Jahrhunderten.* Berlin. Exhibition catalog.

Wehler, Hans-Ulrich. 1987. *Deutsche Gesellschaftsgeschichte: Vom Feudalismus des Alten Reiches bis zur Defensiven Modernisierung der Reformära, 1700–1815.* Vol. 1. Munich.

Wenzel, Georg. 1967. Die Geschichte der Nürnberger Spielzeugindustrie. Dissertation, University of Erlangen-Nuremberg.

Widmann, Hans. 1965. *Der deutsche Buchhandel in Urkunden und Quellen.* 2 vols. Hamburg.

Wild, Reiner. 1987. *Die Vernunft der Väter: Zur Psychographie von Bürgerlichkeit und Aufklärung in Deutschland am Beispiel ihrer Literatur für Kinder.* Germanistische Abhandlungen, no. 61. Stuttgart.

———, ed. 1990. *Geschichte der Deutschen Kinder- und Jugendbuchliteratur.* Stuttgart.

Wittmann, Reinhard. 1991. *Geschichte des deutschen Buchhandels: Ein Überblick.* Munich.

Wünsche, Konrad. 1991. Das Wissen im Bild: Zur Ikonographie des Pädagogischen. *Zeitschrift für Pädagogik,* suppl. 27:273–90.

Yates, Frances. [1966] 1999. *The Art of Memory.* Vol. 3 of *Selected Works.* London.

Zimmermann, Annette. 1981. *Franz von Paula Schrank (1747–1835): Naturforscher zwischen Aufklärung und Romantik.* Neue Münchener Beiträge zur Geschichte der Medizin und Naturwissenschaften, no. 4. Munich.

zur Lippe, Rudolf. 1974. *Naturbeherrschung am Menschen.* 2 vols. Frankfurt am Main.

INDEX

Note: Italicized page numbers indicate figures and plates.